Guillaume-Benjamin Duchenne, George Vivian Poore

Selections from the Clinical Works of Dr. Duchenne (de Boulogne)

Guillaume-Benjamin Duchenne, George Vivian Poore

Selections from the Clinical Works of Dr. Duchenne (de Boulogne)

ISBN/EAN: 9783337277659

Printed in Europe, USA, Canada, Australia, Japan

Cover: Foto ©berggeist007 / pixelio.de

More available books at **www.hansebooks.com**

THE
NATURE AND ELEMENTS OF POETRY

BY

EDMUND CLARENCE STEDMAN
AUTHOR OF "VICTORIAN POETS," "POETS OF AMERICA," ETC.

BOSTON AND NEW YORK
HOUGHTON, MIFFLIN AND COMPANY
The Riverside Press, Cambridge
1892

Copyright, 1892,
By EDMUND CLARENCE STEDMAN.

All rights reserved.

SECOND EDITION.

The Riverside Press, Cambridge, Mass., U.S.A.
Electrotyped and Printed by H. O. Houghton & Co.

TO
LAWRENCE AND FRANCESE TURNBULL,
OF BALTIMORE, MARYLAND,

AND TO THE MEMORY OF THEIR SON,

PERCY GRÆME TURNBULL,

WHOSE LIFE AND DEATH INSPIRED THEIR FOUNDATION
OF THE LECTURESHIP OF POETRY AT JOHNS
HOPKINS UNIVERSITY, THIS VOLUME
IS DEDICATED.

INTRODUCTION

THE series of lectures contained in this volume, although now somewhat revised and extended, formed the initial course, as delivered in 1891, of the Percy Turnbull Memorial Lectureship of Poetry at Johns Hopkins University. In founding that lectureship, Mr. and Mrs. Lawrence Turnbull commemorated the name of their son, Percy Graeme Turnbull, who died in 1887, having nearly completed his ninth year. The brief life of a child, who gave promise of fulfilling the utmost wishes of parents devoted to things good and fair, has been of higher service than that which many whose lights "burn to the socket" are permitted to render.

In conformity with the terms of the gift, a course of lectures is to be delivered annually by some maker or critical student of poetry. There is but one other foundation dedicated to this art alone, as far as I can learn, among British and American universities, that being the chair endowed at Oxford by Henry Birkhead, in 1708, from which much learned argument has been delivered since the time of Warton, and to which we owe the criticism of Doyle, Shairp, Palgrave, and the high discourse of Arnold, in our own day. Had Mr. Lowell's health enabled him to initiate the Turnbull lectureship, the foundation would have derived a lustre at once the light and the despair of his successors. In the shadow of his lamented death it became my duty and distinction to prepare the following lectures, which are now issued in reconsideration of an intention, expressed in my last preceding volume of criticism, to write no more books upon the present theme.

Perhaps it is only natural that such an intention should be overcome by a striking illustration of the fact that, under stress of public neglect or distaste, the lovers of any cause or art find their regard for it more unshaken than ever. It seemed to me a notable thing that at a time when poetry as the utterance of feeling and imagination is strenuously rivalled by other forms of expression, especially by the modern industry of prose fiction; at a time when journalism, criticism, science more than all, not only excite interest, but afford activity and subsistence to original writers; at a time, moreover, when taste is fostered by the wealth of those to whose luxury the architect, the artist, and the musician, rather than the poet, are ready to minister; it seemed to me notable and suggestive that at such a time, though many think of poetry as the voice of the past, a few should still consider it a voice of the future also, and that there should be found what I may call practical idealists, to discover one need of our most liberal schools, and to do this much to relieve it.

I have thought it appropriate that an opening course upon this foundation should relate to the absolute nature of the art which future lecturers will consider more in detail with respect to its technical laws, varied forms, and historic illustrations. These pages, then, treat of the quality and attributes of poetry itself, of its source and efficacy, and of the enduring laws to which its true examples ever are conformed. An attempt to do this within brief limits, notwithstanding the extent of the subject, is not quite impracticable, since whether the "first principles" of any art, even of the philosophy of all arts and knowledge, can be tersely set forth, is not so much in question as is the skill of one who tries to epitomize them.

In the consideration of any subject, however ideal, an agreement as to what shall be denoted by its title may well be established at the outset. Therefore I have not evaded even that which it is so customary to deprecate,—a definition of the thing examined in this treatise. It must be observed that our discussion is of poetry in the concrete, and as the actual record of human expression,— keeping ever in mind, no less, the uncapturable and mysterious spirit from which its energy is derived. I say this, because most essays upon the theme have been produced by one or the other of two classes, — either by transcendentalists who invoke the astral presence but underrate its fair embodiment, or by technical artisans who pay regard to its material guise alone. There is no good reason, I think, why both the essence and the incarnation of poetry may not be considered as directly as those of the less inclusive and more palpable fine arts. At all events, an attempt is made in this volume to do that very thing.

Even this enforced brevity makes it the more needful that my course should be in good faith what its title indicates — elementary. But the simplest laws and constituents, those most patent to common apprehension, are also the most profound and abiding. Their statement must be accurate, first of all; since, as in the present instance, it seeks to determine the initial aim, and a hairbreadth's deviation at the start means a ruinous divergence as the movement progresses. I make no apology, then, for what is elementary and oft-repeated, my wish being, in this opening discussion of that wherewith the Turnbull lectureship is concerned, to derive a statement of first principles from the citation of many illustrious witnesses and creative works. If, therefore, I seem to thresh old straw, it is not without design; and often, instead of making the curious references so easily culled from the less-known books upon our shelves, I repeat passages most famous and familiar,—the more familiar, as a rule, because none apter in illustration can be cited.

In the endeavor to use time to the best advantage, it seemed most feasible to begin with a suggestion of reasons why poetry does not obtain the scientific consideration awarded to material processes, and then to review important outgivings of the past with respect to it; and next, to essay a direct statement of its nature (analyzing the statement logically), and to add a correlative view of its powers and limitations as compared with, and differentiated from, those of the other fine arts. I found it serviceable, afterwards, to divide all poetry—as indeed the product of every art may be divided—into the two main results, creation and self-expression, the vitalities of which are implied in those well-worn metaphysical terms, the objective and the subjective. The former characterization applies to that primitive and heroic song which is the only kind recognized by a Macaulay, with his faculty attuned to the major key. But, after all, there was much self-expression in "the antique," just as there are stately examples of objective creation in the poetry of Christendom. Therefore it was not possible to confine a third lecture entirely to the one, nor a fourth entirely to the other. The creative element, however, is the main topic of the third, while the fourth, entitled "Melancholia," pursues chiefly the stream of self-expression. Together, the two afford all the scope permitted in this scheme for a swift glance at the world's masterpieces. The way now becomes clear for examination of the pure attributes which qualify the art we are considering: — on the side of æsthetics, beauty, — and therewith truth, as concerns the realistic, the instructive, the ethical; then the inventive and illuminating imagination, and passion with its motive power and sacred rage; lastly, the faculty divine, operative through insight, genius, inspiration, and consecrated by the minstrel's faith in law and his sense of a charge laid upon him. A concession from the original scheme appears in the briefness of the section devoted to passion, under which title a poet's emotion should receive the same attention elsewhere given to his taste, sincerity, and imaginative power. My limits compelled me to speak of passion at the opening of the final lecture, where it does not precisely belong, though a necessary excitant of "the faculty divine."

Modern writers upon poetry as an art occupy themselves, as I have hinted, very closely with technical matters,—with "the science of verse," its rhythm, diction, and metrical effects. But these are matters of course for natural poets, each after his own voice and individuality, and technical instruction is obtained by them otherwise than through the schooling which fortifies the practitioners of arts which return subsistence as well as fame. Contenting myself with assuming the need of artistic perfection, I turn to weightier matters of the law, there being no true science of poetry which does not seek after the abstract elements of its power. Nor can any work henceforth be an addition to the literature of the subject, which fails to recognize the obligation of treating it upon scientific lines. For no one now feels the steadfast energy of science more than do the poets themselves, and they realize that, if at first it caused a disenchantment, it now gives promise of an avatar. The readjustment, in truth, is so thoroughly in force that a critic moves with it instinctively. If there is anything novel in this treatise,—anything like construction,—it is the result of an impulse to confront the scientific nature and methods of the thing discussed. Reflecting upon its historic and continuous potency in many phases of life, upon its office as a vehicle of spiritual expression, I have seen that it is only a specific manifestation of that all-pervading force, of which each one possesses a share at his control, and which communicates the feeling and thought of the human soul to its fellows. Thus I am moved to perceive that for its activity it depends, like all other arts, upon Vibrations, — upon ethereal waves conveying impressions of vision and sound to mortal senses, and so to the immortal consciousness whereto those senses minister.

In my opening lecture, I see that mention is made of the disenchantment to which "that airy nothing, the rainbow," has been subjected. But it is precisely because we have discovered its nothingness, — because we know its only being consists in vibrations which impart our sense of light, and of the color scale — that Lippman has been able at last to seize this color scale, and to fix the negative reflecting the light of the eye, the flush of the cheek, to make the sunset eternal, to secure the myriad tints of landscape, — in short, to make a final conquest of nature, and thus to enlarge our basis for the indispensable higher structures of the painter and the poet. Such realism cannot be ignored. It does not lessen ideality; it affords new inspiration. Each time when science fulfils our hope, the poet will be charmed to dream anew, and to impart from his own nature to the semblance of his visions that individuality of tone and form which is the ultimate value of human art.

I have avoided much discussion of schools and fashions. Every race has its own genius, as we say; every period has its own vogues in the higher arts, as well as in those which fashion wholly dominates. There have been "schools" in all ages and centres, but these, it must be acknowledged, figure most laboriously at intervals when the creative faculty seems inactive. The young and ardent, — so long as art has her knight-errantry, so long as there is a brotherhood of youth and hope, — will set out joyously upon their new crusades. Sometimes these are effective, as in the Romantic movement of 1830; but more often, as when observing the neo-romanticists and neo-impressionists, the French and Belgian "symbolists," and just now the "intuitivists," we are taught that, no matter how we reconstruct the altars or pile cassia and frankincense upon them, there will be no mystic illumination unless a flame descends from above. New styles are welcome, but it is a grievous error to believe a new style the one thing needful, or that art can forego a good one, old or new. Our inquiry, then, is concerned with that which never ages, the primal nature of the minstrel's art. Even sturdy thinkers fall into the mistake of believing that a great work loses its power as time goes on. Thus Shakespeare's creations have been pronounced outworn, because he was the last great "poet of feudalism." We might as well say that the truth to human life displayed in Genesis and Exodus, or the synthetic beauty of the Parthenon, or the glory of the Sistine Madonna, will grow ineffective, forgetting that these have the vitality which appertains to the lasting nature of things. No poet can ever outrival Shakespeare, except by a more exceeding insight and utterance. It is well said that great art is always modern, and this is true whether a romantic or a realistic method prevails. Doubtless the prerogative of song is a certain abandonment to the ideal, but this, on the other hand, becomes foolishness unless the real, the truth of earth and nature, is kept somewhere in view. Still, if any artist may be expected to pursue by instinct a romantic method, it is the poet, the very essence of whose gift is a sane ideality. The arbitrary structure of poetry invites us to a region out of the common, and this without danger of certain perils attending the flights of prose romance.

While the poetic drama, for example, *must* be realistic in its truth to life, — first, as to human nature, and, second, in fidelity to the manners of a given time and place, — it shortly fails unless surcharged with romantic passion and ideality. The drama, then, ever catholic and universal, is a standing criticism upon the war of schools, — a war usually foregone whenever the drama reaches and maintains a successful height. I have suggested heretofore the probability that dramatic feeling, and even the production of works in dramatic form, will distinguish the next poetic movement of our own language and haply of this Western world.

But criticism of style and method should be extended to specific productions, and to the writers of a certain period or literature. To the essays which in that wise have come from my own hand this treatise is a natural complement. If inconsistent with them, — if this statement of first principles could not be made up from my books of "applied criticism," I would doubt the integrity of the one and the other; for I have found, in preparing the marginal notes and topical index of the present volume, that nearly every phase and constituent of art has been touched upon, however briefly, which was illustrated in the analytic course of my former essays.

E. C. S.

NEW YORK, *August*, 1892.

CONTENTS

PAGE

I
ORACLES OLD AND NEW 3

II
WHAT IS POETRY? 41

III
CREATION AND SELF-EXPRESSION 75

IV
MELANCHOLIA 111

V
BEAUTY 147

VI

Truth 187

VII

Imagination 225

VIII

The Faculty Divine: Passion, Insight, Genius, Faith 259

INDEX 301

THE NATURE AND ELEMENTS
OF POETRY

THE NATURE AND ELEMENTS OF POETRY

I.

ORACLES OLD AND NEW.

POETRY of late has been termed a force, or mode of force, very much as if it were the heat, or light, or motion known to physics. "The force of heaven-bred poesy." And, in truth, ages before our era of scientific reductions, the *energia* — the vital energy — of the minstrel's song was undisputed. It seems to me, in spite of all we hear about materialism, that the sentiment imparting this energy — the poetic impulse, at least — has seldom been more forceful than at this moment and in this very place.

Our American establishments — our halls of learning and beauty and worship — are founded, as you know, for the most part not by governmental edict; they usually take their being from the sentiment, the ideal impulses, of individuals. Your own institute,[1] still mewing like Milton's eagle its mighty youth, owes its existence to an ideal sentiment, to a

[1] Johns Hopkins University.

most sane poetic impulse, in the spirit of its founder, devoted though he was, through a long and sturdy lifetime, to material pursuits. Its growth must largely depend on the awakening from time to time, in other generous spirits, of a like energy, a similarly constructive imagination.

Amongst all gracious evidences of this ideality thus far calendared, I think of few more noteworthy, of none more beautiful, than those to which we owe the first endowed lectureship of poetry in the United States; the second foundation strictly of its kind, if I mistake not, throughout the universities of the English-speaking world.

The Percy Turnbull Memorial Lectureship:

Whenever a university foundation is established for the study of elemental matters, — of scientific truth or human ideality, — we return to motives from which the antique and the mediæval schools chiefly derived their impulse, if not their constitution. The founders would restore a balance between the arbitrary and the fundamental mode of education. The resulting gain is not the overflow of collegiate resources, not the luxury of learning; not decoration, but enhanced construction. We have a fresh search after the inmost truth of things, the verities of which the Anglo-Florentine songstress was mindful when she averred that poets are your only truth-tellers; of which, also, Lowell, in his soliloquy of "Columbus," was profoundly conscious when he made the discoverer say: —

its fine significance.

> "For I believed the poets; it is they
> Who utter wisdom from the central deep,
> And, listening to the inner flow of things,
> Speak to the age out of eternity."

Within these verities new estates originate; moreover, they perpetually advance the knowledge and methods of the time-honored professions. The present and future influences of a school are more assured when it enters their realm. *The spirit giveth life.* If I did not believe this with my noonday reason and common sense, it would be an imposture for me to discourse to you upon our theme. The sovereign of the arts is the imagination, by whose aid man makes every leap forward; and emotion is its twin, through which come all fine experiences, and all great deeds are achieved. Man, after all, is placed here to live his life. Youth demands its share in every study that can engender a power or a delight. Universities must enhance the use, the joy, the worth of existence. They are institutions both human and humane: not inevitable, except in so far as they become schools for man's advancement and for the conduct of life.

We now have to do with the most ideal and comprehensive of those arts which intensify life and suggest life's highest possibilities. The name of poetry, like that of gentleman, is "soiled with all ignoble use;" but that is because its province is universal, and its government a republic, whose right of franchise any one can exercise without distinction of age, sex, color, or (more's the pity) of morals, brains, or birthright. The more honor, then, to the founders of this lectureship, whose recognition of poetry at its highest is not disturbed by its abuse, and whose munificence erects for it a stately seat among its peers.

Under the present auspices, our own approach can scarcely be too sympathetic, yet none the less free of illusion and alert with a sense of realities. We may well be satisfied to seek for the mere ground-plot of this foundation. I am privileged, indeed, if I can suggest a tentative design for the substructure upon which others are to build and decorate throughout the future of your school. Poetry is not a science, yet a scientific comprehension of any art is possible and essential. Unless we come to certain terms at the outset, if only to facilitate this course, we shall not get on at all.

<small>Design of this treatise.</small>

ENTER the studio of an approved sculptor, a man of genius, and, if you choose, poetic ideality. He is intent upon the model of a human figure, a statue to be costumed in garments that shall both conceal and express the human form. Plainly he has in his mind's eye the outside, the ultimate appearance, of his subject. He is not constructing a manikin, a curious bit of mechanism that imitates the interior — the bones, muscles, arteries, nerves — of the body. He is fashioning the man *as he appears to us*, giving his image the air, the expres-

<small>Tot artes tantæ scientiæ.</small>

sion, of life in action or repose. But you will perceive that even the rude joinery on which he casts his first clay is a structure suggesting a man's interior framework. Ere long the skeleton is built upon; the nude and very man is modeled roughly, yet complete, so that his anatomy shall give the truth, and not a lie, to the finished work. Not until this has been done will the sculptor superadd the drapery — the costume which, be it the symbol of our fall or of our advancement, distinguishes civilized man from the lower animals. At all events, it is a serious risk for the young artist to forego this progressive craftsmanship. Even a painter will rudely outline his figures according to primitive nature before giving them the clothing, which, however full of grace and meaning, is not *themselves*. Otherwise he will be a painter of dead garments, not of soul-possessing men and women. An artist of learning and experience may overleap this process, but only because his hand has become the trained slave of his creative vision, which sees clearly all that can lie beneath.

To the anatomic laws, then, of the human form the sculptor's and the figure-painter's arts _{Preëssentials.} are subservient. The laws of every art are just as determinate, even those pertaining to the evasive, yet all-embracing art of poesy, whose spirit calls other arts to its aid and will imitate them, as art itself imitates nature; which has, in truth, its specific method and also the reflex of all other methods. I do not speak of the science, even of the art, of verse. Yet to know the spirit of poetry we must observe, with the temper of philosophers, its preëssentials in the concrete. Even its form and its method of work must be recognized as things of dignity: the material symbols and counterparts, as in Swedenborg's cosmos, of the spirit which is reality.

And thus, I say, we must obtain at least a serviceable definition of the word poetry for our present use. In beginning this course, it is well to let the mists rise, at least to have none of our own brewing. The sentimentalists invariably have befogged our topic. I ask you to divest your minds, for the moment, of sentimentalism, even of sentiment, and to assume, in Taine's phrase, that we are to begin by realizing "not an ode, but a law." Applied criticism — that which regards specific poets and poems — is a subsequent affair. Let us seek the generic elements that are to govern criticism by discovering and applying its standards. If you ask, To what end? I reply, That we may avoid dilettanteism. We are not a group of working artists, but they possess something we can share; to wit, the sincere and even ascetic mood that wishes no illusions and demands a working basis. But again, to what purpose? Surely not for the development of a breed of poets! Consider the tenuous voices of minnesingers far and near, whose music rises like the chirping of locusts by noonday and of meadow-frogs at night. Each has his fault-

A working basis needed;

but not for the promotion of versifying.

less little note, and while the seasonable chorus blends, it is humored by some and endured by most, quite as a matter of course, and the world goes on as usual. Human suffering may have been greater when the rhapsodist flourished and printing was unknown, when one was waylaid at the corners of the market-place, and there was no escape but in flight or assassination. And if our object were to train poets, and a past-master were on the rostrum, his teachings would be futile unless nature reassorted her averages. Fourier accounted for one poet in his phalanstery of a thousand souls; yet a shrewder estimate would allow but one memorable poet to a thousand phalansteries, in spite of the fact that even nature suspends her rules in countenance of youth's prerogative, and unfailingly supplies a laureate for every college class. With respect to training, the catalogues term a painter the pupil of Bonnat, of Duran, of Cabanel; a musician, pupil of Rubinstein or Liszt. But the poet studies in his own atelier. He is not made, his poetry is not made, by *a priori* rules, any more than a language is made by the grammarians and philologists, whose true function is simply to report it.

Nature both makes and trains a poet.

Yet even the poet has his teachers: first of all, since poetry is vocal, those from whom he learns the speech wherewith he lisps in numbers. In the nursery, or on the playground, he is as much at school as any young artist taking his initial lessons in the drawing-class, or a young singer put to his first exercises. Later on, he surely finds his way to the higher gymnasium; he reads with wonder and assimilation the books of the poets. Thus not only his early methods, but his life-long expression, his vocabulary, his confines and liberties, will depend much upon early associations, and upon the qualities of the models which chance sets within his way. As to technical ability, what he seeks to acquire after the formative period relatively counts for little; his gain must come, and by a noble paradox, from learning to unlearn, from self-development; otherwise his utterance will never be a force. One poet's early song, for example, has closely echoed Keats; another's, Tennyson; afterward, each has given us a motive and a method of his own, yet he was first as much a pupil of an admirable teacher as those widely differing artists, Couture and Millet, were pupils of Delaroche. Still another began with the Italian poets, and this by a fortunate chance,—or rather, let us say, by that mysterious law which decrees that genius shall find its own natural sustenance. In time he developed his own artistic and highly original note, with a spirituality confirmed by that early pupilage.

I assume, then, that the poet's technical modes, even the general structure of a masterwork, come by intuition, environment, reading, experience; and that *The natural method.* too studious consideration of them may perchance retard him. I suspect that no instinctive poet bothers himself about such matters in advance; he doubtless casts his work in the form and measures that come with its thought to him, though he afterward may pick up his dropped feet or syllables at pleasure. If he ponders on the Iambic Trimeter Catalectic, or any of its kin, his case is hopeless. In fact, I never have known a natural poet who did not compose by ear, as we say: and this is no bad test of spontaneity. And as for rules,—such, for example, as the Greeks laid down,—their efficacy is fairly hit off in that famous epigram of the Prince de Condé, when the Abbé d'Aubignac boasted that he closely observed the rules of Aristotle: "I do not quarrel with the Abbé d'Aubignac for having so closely followed the precepts of Aristotle; but I cannot pardon the precepts of Aristotle that occasioned the Abbé d'Aubignac to write so wretched a tragedy." We do see that persons of cleverness and taste learn to write agreeable verses; but the one receipt for making a poet is in the safe-keeping of nature and the foreordaining stars.

On the other hand, the mature poet, and no less the lover of poetry, may profitably observe what secrets of nature are applied to lyri- *One end in view.* cal creation. The first Creator rested after his work, and saw that it was good. It is well for an artist to study the past, to learn what can be done and what cannot be done acceptably. A humble music-master can teach a genius not to waste his time in movements proved to be false. Much of what is good is established, but the range of the good is infinite;

that which is bad is easily known. If there be a mute and to-be-glorious Milton here, so much the better. And for all of us, I should think, there can be no choicer quest, and none more refining, than, with the Muse before us, to seek the very well-spring and to discover the processes of her "wisdom married to immortal verse."

Artistic reserve. We owe to the artist's feeling that his gift is innate, and that it does produce "an illusion on the eye of the mind" which, he fears, too curious analysis may dispel: to this we doubtless owe his general reluctance to talk with definiteness concerning his art. Often you may as well ask a Turk after his family, or a Hindu priest concerning his inner shrine. I have put to several minstrels the direct question, "What is poetry?" without obtaining a categorical reply. One of them, indeed, said, "I can't tell you just now, but if you need a first-class example of it, I'll refer you to my volume of 'Lyrics and Madrigals.'" But when they do give us chips from their workshop, — the table-talk of poets, the stray sentences in their letters, — these, like the studio-hints of masters, are both curt and precious, and emphatically refute Macaulay's statement that good poets are bad critics. They incline us rather to believe with Shenstone that "every good poet includes a critic; the reverse" (as he added) "will not hold."

Even a layman shares the artist's hesitation to

discourse upon that which pertains to human emotion. Because sensation and its causes are universal, the feeling that creates poetry for an expression, and the expression itself, in turn exciting feeling in the listener, are factors which we shrink from reducing to terms. An instinctive delicacy is founded in nature.. To overcome it is like laying hands upon the sacred ark. One must be assured that this is done on the right occasion, and that, at least for the moment, he has a special dispensation. A false handling cheapens the value of an art — puts out of sight, with the banishment of its reserve, what it might be worth to us. All have access to the universal elements; they cost nothing, are at the public service, and even children and witlings can toy with and dabble in them. So it is with music, poetry, and other general expressions of feeling. Most people can sing a little, any boy can whistle — and latterly, I believe, any girl who would defy augury, and be in the fashion. Three fourths of the minor verse afloat in periodicals or issued in pretty volumes corresponds to the poetry of high feeling and imagination somewhat as a boy's whistling to a ravishing cavatina on the Boehm flute. As a further instance, a knack of modelling comes by nature. If sculptor's clay were in every road-bank, and casts from the antique as common as school readers and printed books of the poets, we probably should have reputed Michelangelos and Canovas in every village instead of here and there a Ward, a St. Gaudens, or a Donoghue.

<small>The heritage of all,</small>

But it is precisely the arts in which anybody can dabble that the elect raise to heights of dignity and beauty. Those who realize this indulge a pardonable foible if they desire to reserve, like the Egyptian priests, certain mysteries, if only *pro magnifico*. Besides, there are periods when the utility of artistic analysis is not readily accepted by those who make opinion. Economics and sociology, for example, largely absorb the interest of one of our most scholarly journals. Its literary and art columns are ably conducted. The chief editor, however, told me that he knew little of æsthetics, and cared to know less; and in such a way as to warrant an inference that, though well disposed, he looked upon art and song and poetry very much as Black Bothwell regarded clerkly pursuits, — that they were to him what Italian music seemed to Dr. Johnson, in whose honest eyes its practitioners were but fiddlers and dancing-masters. This undervaluation by a very clever man is partly caused, if not justified, one must believe, by the vulgarization of the arts of beauty and design. Yet these arts belong as much to the order of things, and indirectly make as much for wealth, as the science of economics, and they make as much for social happiness as the science of sociology, — if, indeed, they are to be excluded from either.

the crown of few.

A traditional undervaluation.

Can we, even here, take up poetry as a botanist

takes up a flower, and analyze its components? Can
we make visible the ichor of its proto- Can poetry be defined?
plasm, and recognize a something that
imparts to it transcendency, the spirit of the poet
within his uttered work? Why has the question
before us been so difficult to answer? Simply because
it relates to that which is at once inclusive
and evasive. There is no doubt what sculpture and
painting and music and architecture seem to be;
the statements of critics may differ, but the work
is visible and understood. Do you say with the
philosophers that poetry is a sensation, that its quality
lies in the mind of the recipient, and hence is
indefinite? The assertion applies no less to the
plastic arts and to music, yet the things by which
those excite our sensations are well defined, and
what I seek is the analogous definition of the spoken
art. It has been said that "one element must forever
elude researches, and that is the very element
by which poetry is poetry." I confess we cannot
define the specific perfume of a flower; but there is
a logical probability that this conveys itself alike to
all of us, that the race is as but one soul in receiving
the impression. I think we can seize upon all other
conditions that make a flower a flower or a poem a
poem.

Edgar Poe avowed his belief in the power of
words to express all human ideas, — a Whether language is inadequate.
belief entertained by Joubert also. Nor
have I any doubt that for every clear thought, even

for every emotion, words have been, or can be, found, as surely as there is a conquest of matter by the spirit; that speech, the soul's utterance, shares the subtilties of its master. Where it seems to fail, the fault is in the speaker. As a race goes on, both its conceptions and its emotions are clearer and richer, and language keeps pace with them. The time may come, indeed, when thought will not be "deeper than all speech," nor "feeling deeper than all thought." If we still lag in emotional expression, we can excite feelings similar to our own by the spells of art. I do not see why the primary elements of poetry in the concrete should not be stated without sophistication, and as clearly as those of painting, music, or architecture. They have, in fact,

<small>Oracles, old and new.</small> been stated fragmentarily by one and another poet and thinker, most of whom agree on certain points. True criticism does not discredit old discovery in its quest for something more. Its office, as Mill says of philosophy, is not to set aside old definitions, but it "corrects and regulates them." It does not differ for the sake of novelty, but formulates what is, and shall be, of melody and thought and feeling, and what no less has been since first the morning stars sang together. I must ask you, then, to permit me, in this opening lecture, very swiftly to review familiar and historic utterances, from which we may combine principles eminently established, and, if need be, to add some newly stated factor, in our subsequent effort to for-

mulate a definition of poetry that shall be scientifically clear and comprehensive, and also to establish limits beyond which speculation is foreign to the design of this lecture-course.

Various poets and thinkers, each after his kind, have contributed to such a definition. I have mentioned Aristotle. He at least applied to the subject a cool and level intellect; and his formula, to which in certain essentials all must pay respect, is an ultimate deduction from the antique. It fails of his master Plato's spirituality, but excels in precision. Aristotle regards poetry as a structure whose office is imitation through imagery, and its end delight, — the latter caused not by the imitation, but through workmanship, harmony, and rhythm. The historian shows what has happened, the poet such things as might have been, devoted to universal truth rather than to particulars. The poet — the ποιητής — is, of course, a maker, and his task is invention. Finally, he must feel strongly what he writes. Here we have the classical view. The Greeks, looking upon poetry as a fine art, had no hesitation in giving it outline and law. _{The antique or classical idea.} _{From Aristotle to Goethe.}

Naturally an artist like Horace assented to this conception. Within his range there is no more enduring poet; yet he excludes himself from the title, and this because of the very elements which make him so modern, — his lyrical _{Horace, Dryden, and others.}

grace and personal note. With Aristotle, he yielded the laurel solely to heroic dramatists and epic bards. His example is followed by our brave old Chapman, Homer's bold translator, who declares that the *energia* of poets lies in "high and hearty Invention." Dryden also accepts the canon of Imitation, but avows that "Imaging is, in itself, the height and life of it," and cites Longinus, for whom poetry was "a discourse which, by a kind of enthusiasm, or extraordinary emotion of the soul, makes it seem to us that we behold those things which the poet paints." Landor, the modern Greek, whose art was his religion, repeats that "all the imitative arts have delight for their principal object; the first of these is poetry; the highest of poetry is the tragic." But recognition of only the structure of verse, without its soul, deadened the poetry of France in her pseudo-classical period, from Boileau to Hugo, so that it could be declared, as late as A. D. 1838, that "in French literature that part is most poetry which is written in prose." Even the universal Goethe repressed his "noble rage" by the conception of poetry as an art alone, so that Heine, a pagan of the lyrical rather than of the inventive cast, said that this was the reason why Goethe's work did not, like the lesser but more human Schiller's, "beget deeds." "This is the curse," he declared, "of all that has originated in mere art." Like Pygmalion and the statue, "his kisses warmed her into life,

As an art alone.

but, so far as we know, she never bore children."[1] Goethe's pupil, the young Matthew Arnold, accepted without reserve the antique notion of poetry. "Actions, human actions," he cried, "are the eternal objects of the muse." In after years, as we shall see, he formed a more sympathetic conception.

Other poets have thrown different and priceless alloys into the crucible from which is to flow the metal of our seeking, adding fire and sweetness to its tone. The chiefs of the romantic movement, so near our own time, believed Passion to be the one thing needful. Byron was its fervent exemplar. In certain moods, it is true, he affected to think that he and his compeers were upon a wrong system, and he extolled the genius and style of Pope. But this was after all had got the seed of his own flower. It was plainly an affectation of revolt from his own affectation, with haply some prophetic sense of naturalism as a basis for genuine emotion. His summing up is given in "Don Juan":— *The Romantic view. Poetry as the lyrical expression of Emotion.* *Byron.*

> "Thus to their extreme verge the passions brought
> Dash into poetry, which is but passion,
> Or at least was so, ere it grew a fashion."

Moore, light-weight as he was, aptly stated the Byronic creed: "Poetry ought only to be employed as an interpreter of feeling." This is certainly true, as far as it goes, and agrees with Mill's

---

[1] But see Ovid, *Met.* x. 297:—
 "Illa Paphon genuit, de quo tenet insula nomen."

later but still limited canon, that poetry is emotion expressed in lyrical language.[1] But a complete definition distinguishes the thing defined from everything else; it denotes, as you know, "the species, the whole species, and nothing but the species." Bascom and Ruskin follow Mill, but Ruskin adds other elements, saying that poetry is the suggestion, by the "imagination," of noble "thoughts" for noble emotions. This does not exclude painting and other emotional and imaginative arts. In truth, he is simply defining art, and takes poetry, as Plato might, as a synonym for art in all its forms of expression.

<small>Mill and Ruskin.</small>

An elevated view, on the whole, is gained by those who recognize more sensibly the force of Imagination. Here the twin contemplative seers, Wordsworth and Coleridge, lift their torches, dispersing many mists. They saw that poetry is not opposed to prose, of which verse is the true antithesis, but that in spirit and action it is the reverse of science or matter of fact. Imagination is its polestar, its utterance the echo of man and nature. The poet has no restriction beyond the duty of giving pleasure. Nothing else stands between him and the very image of nature, from which a hundred barriers shut off the biographer and historian. Wordsworth admits the need of emotion, but renounces taste. Coleridge plainly has the instinct for beauty and the spell of measured words.

<small>Imagination.</small>

<small>The Lake School.</small>

---

[1] J. S. Mill's *Thoughts on Poetry and its Varieties*, 1833.

The chief contributions of the Lake School to our definition are the recognition of the imagination and the antithesis of science to poetry.[1] The pessimist Schopenhauer, who wrote like a musician on music, like a poet on poetry, yet with wholly impassive judgment, also avows that poetry is "the art of exciting by words the power of the imagination," and that it must "show by example what life and the world are."

From the attributes of invention, passion, and imagination may perhaps be deduced what seems to others the specific quality of the poet, the very quintessence of his gift. *The Platonic conception.* What should I mean, save that which Aristotle's master considered the element productive of all others and a direct endowment from heaven, — *Inspiration.* the Inspiration governing creative, impassioned, imaginative art? The poet's soul was, according to Plato, in harmonic relation with the soul of the universe. It is true that in the "Republic" he supplies Aristotle with a technical basis; *Plato, in "The Republic."* furthermore, as an idealist playing at government, he is more sternly utilitarian than even the man of affairs. The epic and dramatic makers of "imitative history" are falsifiers, dangerous for their divine power of exciting the passions and unsettling the minds of ordinary folk. He admires a poet, and would even crown him, but feels bound to escort

---

[1] Coleridge's *Introductory Matter on Poetry, the Drama, and the Stage;* Wordsworth's Prefaces and Appendix to *Lyrical Ballads,* etc.

him to the side of the ship Republic and to drop him overboard, as the Quaker repulsed the boarder, with the remark, "Friend, thee has no business here!" But this is Plato defying his natal goddess in a passing ascetic mood; Plato, in whose self the poet and philosopher were one indeed, having ever since been trying, like the two parts of his archetypal man, to find again so perfect a union. In his more general mood he atones for such wantonness, reiterating again and again that the poet is a seer, possessed of all secrets and guided by an inspiring spirit; that without his second sight, his interpretation of the divine ideas symbolized by substance and action, his mission would be fruitless.

Those who take this higher view revere the name of Plato, though sometimes looking beyond him to the more eastern East, whence such occult wisdom is believed to flow,— to such sayings as that ascribed to Zoroaster,[1] "Poets are standing transporters; their employment consists in speaking to the Father and to Matter, in provoking apparent copies of unapparent natures, and thus inscribing things unapparent in the apparent fabric of the world."

From Plato to Emerson.

Cicero, deeply read in Plato, could not conceive of a poet's producing verse of grand import and perfect rhythm without some heavenly inbreathing of the mind. The soul's highest prerogative was to contemplate the order of celestial things

Cicero.

[1] Cited by F. B. Sanborn in a paper on Emerson.

and to reproduce it. Transcendental thinkers — such as Lord Bacon in his finest vein — recognize this as its office. While Bacon's general <small>Bacon.</small> view of poetry is that all "Feigned History" (as he terms it), prose or verse, may be so classed, he says the use of it "hath been to give some shadow of satisfaction to the mind of man in those points wherein the nature of things doth deny it"; and again, that it is thought to "have some participation of divineness because it doth raise and erect the mind, by submitting the shews of things to the desires of the mind." Sidney's flawless <small>Sidney.</small> "Defense of Poesie"[1] exalts the prophetic gift of the vates above all art and invention. In our day Carlyle clung to the supremacy of inspiration, in art no less than in action. But no one since Plotinus has made it so veritably the golden dome of the temple as our seer of seers, Emerson, in whose belief the artist does not create so much as report. The soul works through him. "Poetry is the perpetual endeavor to express the spirit of the thing." And thus all the Concord group, notably <small>The Concord</small> Dr. W. T. Harris, in whose treatises of <small>School.</small> Dante and other poets the spiritual interpreting

---

[1] Prof. Albert S. Cook, in his edition of Sidney's tractate, remarks concerning the title: "The *Defense* was not published till 1595, and then by two different printers, Olney and Ponsonby. The former gave it the title, *An Apologie for Poetrie;* the latter, *The Defence of Poesie.* It is doubtful which of these appeared the earlier. . . . Sidney himself refers to the treatise as 'a pitiful defense of poor poetry.'"

power of the bard is made preëminent. The subtlest modern poet of life and thought, Browning, <small>Browning.</small> has left us only one prose statement of his art, but that is the lion's progeny. The poet's effort he saw to be "a presentment of the correspondency of the universe to the Deity, of the natural to the spiritual, and of the actual to the ideal." Spiritual progress, rather than art, is the essential thing. A similarly extreme view led Carlyle (himself, like <small>Carlyle.</small> Plato, a poet throughout) to discountenance the making of poetry as an art. Carried too far, the Platonic idea often has vitiated the work <small>Transcendental strength and weakness.</small> of those minor transcendentalists who reduce their poetics to didactics, and inject the drop of prose that precipitates their rarest elixir. Their creed, however,— with its inclusion of the bard as a revealer of the secret of things, — while not fully defining poetry, lays stress upon its highest attribute.

Thus we see that many have not cared to speak <small>Per ambages, ut mos oraculis.</small> absolutely, and more have failed to discriminate between the thing done and the means of doing. Poetry is made a Brahma, at once the slayer and the slain. A vulgar delusion, that of poetasters, is to confound the art with its materials. The nobler error recognizes the poetic spirit, but not that spirit incarnate of its own will in particular and concrete form. The outcome is scarcely more exact and substantial than the pretty thesis caroled by "one of America's pet Marjories" in her tenth year, and long since become of record. This child's heart detected "poetry, poetry everywhere!" and proclaimed that

> " You breathe it in the summer air,
> You see it in the green wild woods,
> It nestles in the first spring buds.
> . . . .
> 'T is poetry, poetry everywhere —
> It nestles in the violets fair,
> It peeps out in the first spring grass —
> Things without poetry are very scârce."

That our naïve little rhymer was a sibyl, and her statement hardly more vague than the definitions of poetry offered by older philosophers, who will deny?

All in all, various writers connected with the art movement of the present century have most sensibly discussed the topic. They recognize poetry as an entity, subject to expressed conditions. Hazlitt and Leigh Hunt logically distinguished between it and poetic feeling, and believed one to be the involuntary utterance of the other, sympathetically modulating the poet's voice to its key. Shelley, the Ariel of songsters, came right down to the ground of our enchanted isle, laying stress upon the dependence of the utterance on rhythm and order — on "those arrangements of language, and especially metrical language, which are created by that imperial faculty whose throne is contained within the invisible nature of man." More recently the poet-critic, Theodore Watts, in the best modern essay upon the

Clearer statements.

Shelley's noble "Defence of Poetry," 1821.

subject,[1] says that "absolute poetry is the concrete and artistic expression of the human mind in emotional and rhythmical language." Here we certainly are getting out of the mists. In these formulas an effort for precision is apparent, and the latest one would be satisfactory did it insist more definitely, within itself, upon the office of the imagination, and upon the interpretative gift which is the very soul of our art.

The ideas presented by many of the poets seem in the main conformed to their own respective gifts, and therefore in a sense limited. Thus, years after Schlegel had termed poetry "the power of creating what is beautiful, and representing it to the eye or ear," our disciple of taste, Poe, who avowed that poetry had been to him "not a purpose, but a passion," amended Schlegel's terms with the adjective needed to complete his own definition — "the *Rhythmical* Creation of Beauty." Never did a wayward romancer speak with a sincerer honesty of the lyrical art, and he clenched his statement by adding that its sole arbiter was Taste. If you accept beauty in a comprehensive sense, including all emotions, truths, and ethics, accept this definition as precise and unflinching. But Poe confines its meaning to the domain of æsthetics, which of itself he thought opposed to vice on account of her deformity; furthermore, he restricts it to what he terms supernal beauty, the

The personal limitation.

Lecture on "The Poetic Principle," 1845.

[1] "Poetry," in the *Encyclopædia Britannica*, Ninth Edition.

note of sadness and regret. This was simply his own highest range and emotion. His formula, however, will always be tenderly regarded by refined souls, for Beauty, pure and simple, is the alma mater of the artist; her unswerving devotee is absolved — many sins are forgiven to him who has loved her much.

But often a poet, great or small, has burnished some facet of the jewel we are setting. Milton's declaration that poetry is "simple, sensuous, passionate," is a recognition of its most effective attributes.[1] Lowell has sprinkled the whole subject with diamond-dust, and he, of all, perhaps could best have given a new report of its tricksy spirit. Arnold's phrase, "a criticism of life, under the conditions fixed for such a criticism by the laws of poetic truth and poetic

The Miltonic canon.

Arnold's Delphic outgiving.

---

[1] Milton's phrase has become familiar as a proverb since Coleridge used it with great force in the prelude to his lectures on Shakespeare and on the Drama, but it is seldom quoted with its context, as found in the tractate *On Education*, addressed to Samuel Hartlib, A. D. 1644. The poet there speaks of "Rhetoric" as an art "to which poetry would be made subsequent, or indeed rather precedent, as being less subtile and fine, but more simple, sensuous, and passionate. I mean not here the prosody of a verse, which they could not but have hit on before among the rudiments of grammar; but that sublime art which in Aristotle's poetics, in Horace, . . . teaches what the laws are of a true epic poem, what of a dramatic, what of a lyric, what decorum is, which is the grand masterpiece to observe. This would make them soon perceive what despicable creatures our common rhymers and playwriters be; and show them what religious, what glorious and magnificent, use might be made of poetry, both in divine and human things."

beauty," is of value, yet one of those definitions which themselves need a good deal of defining. With the exception of Mr. Watts, we see that not even the writers of our logical period have condensed into a single clause a statement that establishes, practically and inclusively, the basis on which our art sustains its enrapturing vitality, and Mr. Watts's statement leaves something for inference and his after-explanation. Before endeavoring, in the next lecture, to construct a framework that may serve our temporary needs, I wish to consider briefly the most suggestive addition which this century has made to the elements previously observed. I refer to the assertion of Wordsworth and Coleridge that poetry is "the antithesis to science."

What does this assertion mean, and how far does its bearing extend? The poet has two functions, one directly opposed to that of the scientist, and avoided by him, while of the other the scientist is not always master. The first is that of treating nature and life as they *seem*, rather than as they are; of depicting phenomena, which often are not actualities. I refer to physical actualities, of which the investigator gives the scientific *facts*, the poet the *semblances* known to eye, ear, and touch. The poet's other function is the exercise of an insight which pierces to spiritual actualities, to the meaning of phenomena, and to the relations of all this scientific knowledge.

<small>Poetry as the antithesis to science.</small>

To illustrate the distinction between a poet's, or other artist's, old-style treatment of things as they seem and the philosopher's statement of them as they are, I once used an extreme, and therefore a serviceable, example; to wit, the grand Aurora fresco in the Rospigliosi palace. Here you have the childlike, artistic, and phenomenal conception of the antique poets. To them the Dawn was a joyous heroic goddess, speeding her chariot in advance of the sun-god along the clouds, while the beauteous Hours lackeying her scattered many-hued blossoms down the eastern sky. For the educated modern there is neither Aurora nor Apollo; there are no winged Hours, no flowers of diverse hues. His sun is an incandescent material sphere, alive with magnetic forces, engirt with hydrogenous flame, and made up of constituents more or less recognizable through spectrum analysis. The colors of the auroral dawn — for the poet still fondly calls it auroral — are rays from this measurable incandescence, refracted by the atmosphere and clouds, under the known conditions that have likewise put to test both the pagan and biblical legends of that prismatic nothing, the rainbow itself.[1] The stately blank-verse poem, "Orion," which the late Hengist Horne published at a farthing half a century ago, is doubtless our most imaginative rendering of the legend which

The Real and the Apparent.

---

[1] "There was an awful rainbow once in heaven:
We know her woof, her texture; she is given
In the dull catalogue of common things."
      Keats: *Lamia.*

placed the blind giant in the skies. The most superb of constellations represents even in modern poetry a mythical demigod. In science it was but the other day that the awful whirl of nebulæ developed by the Lick telescope revealed it to us almost as a distinct universe in itself.

But to show the distinction as directly affecting modes of expression, take the first of countless illustrations that come to hand ; for instance, the methods applied to the treatment of one of our recurrent coast storms. The poet says :—

<small>A modern instance.</small>

> "When descends on the Atlantic
>   The gigantic
> Storm-wind of the Equinox,
> Landward in his wrath he scourges
>   The toiling surges
> Laden with sea-weed from the rocks."

Or take this stanza by a later balladist :—

> "The East Wind gathered, all unknown,
>   A thick sea-cloud his course before :
> He left by night the frozen zone,
>   And smote the cliffs of Labrador ;
> He lashed the coasts on either hand,
> And betwixt the Cape and Newfoundland
>   Into the bay his armies pour."

All this impersonation and fancy is translated by the Weather Bureau into something like the following :—

An area of extreme low pressure is rapidly moving up the Atlantic coast, with wind and rain. Storm-centre now off Charleston, S. C. Wind N. E. Velocity, 54. Barometer, 29.6. The disturbance will reach New York on Wednesday, and proceed eastward to the Banks and Bay St. Lawrence. Danger-signals ordered for all North Atlantic ports.

We cannot too clearly understand the difference between artistic vision and scientific analysis. The poet in his language and the painter with his brush are insensibly and rightly affected by the latter. The draughtsman, it is plain, must depict nature and life as they seem to the eye, and he needs only a flat surface. The camera has proved this, demonstrating the fidelity in outline and shadow of drawings antedating its use. The infant, the blind man suddenly given sight, see things in the flat as we do, but without our acquired sense of facts indicated by their perspective. We have learned, and experience has trained our senses to instant perception, that things have the third dimension, that of thickness, and are not equally near or far. The Japanese, with an instinct beyond that of some of his Mongolian neighbors, avoids an extreme flat treatment by confining himself largely to the essential lines of objects, allowing one's imagination to supply the rest. He carries suggestiveness, the poet's and the artist's effective ally, to the utmost. Still, as Mr. Wores says, he has no scruples about facts, "for he does not pretend to draw things as they are, or should be, but as they seem." Now, it is probable that the

<small>The distinction chiefly one of methods.</small>

Aryan artist is born with a more analytic vision than that of the Orient; if not, that he does instinctively resist certain inclinations to draw lines just as they appear to him. But this natural resistance unquestionably was long ago reinforced by his study of the laws of perspective. The generally truer and more effective rendering of outline and shadow by Western masters cannot be denied, and furnishes an example of the aid which scientific analysis can render to the artist. In just the same way, we may see, empirical knowledge is steadily becoming a part of the poet's equipment, and, I have no doubt, is by inherited transmission giving him at birth an ability to receive from phenomena more scientifically correct impressions. For his purposes, nevertheless, the portrayal of things as they seem conveys a truth just as important as that other truth which the man of analysis and demonstration imparts to the intellect. It is the *methods* that are antithetical.

The poet's other function, which the scientist does not avoid, but which research alone does not confer upon him, is that of seizing the abstract truth of things whether observed or discovered. It has been given out, though I do not vouch for it, that Edison obtains some of his ideas for practical invention from the airy flights of imagination taken by writers of fiction. In any case, it is clear that with respect to inventive surmise the poet is in advance: the investigator, if he would

Discovery through Imagination.

leap to greater discoveries, must have the poetic insight and imagination, — be, in a sense, a poet himself, and exchange the mask and gloves of the alchemist for the soothsayer's wand and mantle. Those of our geologists, biologists, mechanicians, who are not thus poets in spite of themselves must sit below the seers who by intuition strike the trail of new discovery. For beyond both the phantasmal look of things and full scientific attainment there is a universal coherence — there are infinite meanings — which the poet has the gift to see, and by the revelation and prophecy of which he illumines whatever is cognizable.

The so-called conflict of science and religion, in reality one of fact and dogma, has been waged obviously since the time of Galileo. Its annals are recorded. It was the sooner inevitable because science takes nothing on faith. The slower, but equally prognosticable, effect of exact science on poetry, though foreseen by the Lake School, was not extreme until recently, — so recently, in fact, that a chapter which I devoted to it in 1874 was almost the first extended consideration that it received. [Cp. "Victorian Poets": pp. 7-21.] Since then it has been constantly debated, and not always radically. That the poets went on so long in the old way, very much like the people who came after the deluge, was due to two conditions. First, their method was so ingrained in literature, so common to the educated world, that it sustained a beauteous phantasmagory against all odds. Again, the poets have walked in lowly ways, and each by himself; they have no proud temporal league and station, like the churchmen's, to make them timid of innovation, of any new force that may shake their roof-trees. They have been gipsies, owning nothing, yet possessed of everything without the care of it. At last they see this usufruct denied them; they are bidden to surrender even their myths and fallacies and inspiring illusions. With a grace that might earlier have been displayed by the theologians, they are striving to adapt art to its conditions, though at the best it is a slow process to bring their clientage to the new ideality. Though the imagery and diction which have served their use, and are now absurd, must cease, the creation of something truer and nobler is not the work of a day, and of a leader, but of generations. So there is a present struggle, and the poets are sharing the discomfort of the dogmatists. The forced marches of knowledge in this age do insensibly perturb them, even give the world a distaste for a product which, it fears, we must distrust. The new learning is so radiant, so novel, and therefore seemingly remarkable, that of itself it satiates the world's imagination. Even the abashed idealists, though inspired by it, feel it becoming to fall into the background. Some of them recognize it with stoical cynicism and stern effect. In Balzac's "The Search for the Absolute," Balthazar's wife, suffering agonies, makes an attempt to dis-

Through night to light.

suade him from utterly sacrificing his fortune, his good name, even herself, in the effort to manufacture diamonds. He tenderly grasps her in his arms, and her beautiful eyes are filled with tears. The infatuated chemist, wandering at once, exclaims: "Tears! I have decomposed them: they contain a little phosphate of lime, chloride of sodium, mucin, and water." Such is the last infirmity of noble minds to-day.

We latterly find our bards alive to scientific revelations. It has been well said that a "Paradise Lost" could not be written in this century, even by a Milton. In his time the Copernican system was acknowledged, but the old theory of the universe haunted literature and was serviceable for that conception of "man's first disobedience," and the array of infernal and celestial hosts, to which the great epic was devoted. In our own time such a poet as Tennyson, to whom the facts of nature are everything, does not make a lover say, "O god of day!" but Cp. "Poets of America": pp. 153-155, 262, 382.

> "Move eastward, happy earth, and leave
> Yon orange sunset, waning slow."

Browning, Banville, Whitman, Emerson earliest and most serenely,— in fact, all modern intellectual poets, — not only adapt their works to physical knowledge, but, as I say, often forestall it. Even as we find them turned savants, we find our Clerk Maxwells, Roods, Lodges, Rowlands, poets in their quick guesses and assump- Scientific intuitions.

tions. Imaginative genius is such that often one of its electric flames will come through what is ordinarily a non-conductor. That term, howbeit, cannot be applied to an American scientist[1] who enjoys the distinction of being at once a master of abstruse mathematics and a brilliant writer of very poetic novels, and to whom I put the same question I have addressed to poets, — simply, What is poetry? He Letter from an imaginative savant. repaid me with a letter setting forth in aptest phrase his own belief in the kindred imaginations of the physicist and the poet. Naturally he considers the physical discoverer just now more triumphant and essential. "His study," he says, "is relations. When he cannot discover them, he invents them, — strings his fact-beads on the thread of hypothesis." After some illustrations of this, he sets present research above past fancy, and exclaims: "Compare the wings of light on which we ascend with a speed to girdle the earth eight times a second, to sift the constitution of stars, with the steed of Mohammed and its five-league steps and eyes of jacinth! What a chapter the Oriental poet could give us to-day in a last edition of Job — founding the conception of the Unknown on what we *know* of his works, instead of on our ignorance of them. I want a new Paul to rewrite and restate the doctrine of immortality."

But here the poet may justly break in and say, It is not from investigators, but from the divine

[1] Prof. A. S. Hardy, of Dartmouth College.

preachers, that we inherit this doctrine of immortality. Being poets, through insight they saw it to be true, and announced it as revealed to them. *(Insight first of all.)* Let science demonstrate it, as it yet may, and the idealists will soon adjust their imagery and diction to the resulting conditions. It is only thus they *can* give satisfaction and hold their ground. The prolongation of worn-out fancy has been somewhat their own fault, and it is just they should suffer for it. Still, although we may shift externals, the idealists will be potent as ever; their strength lies not in their method, but in their sovereign perception of the relations of things. Even the theologians no longer dismiss facts with the quotation, "Canst thou by searching find out God?" The world has learned that at all events we can steadily broaden and heighten our conception of him. We are beginning to verify Lowell's prophetic statement: —

> "Science was faith once; Faith were science now
> Would she but lay her bow and arrows by
> And arm her with the weapons of the time."

Theology, teaching immortality, now finds science deducing the progressive existence of the soul as an inference from the law of evolution. *(Aspects of the transition.)* Poetry finds science offering it fresh discovery as the terrace from which to essay new flights. While realizing this aid, a temporary disenchantment is observed. The public imagination is so intent upon the marvels of force, life, psychology, that it concerns itself less with the poet's ideals. Who cares for the ode pronounced at the entrance of this Exposition, while impatient to reach the exhibits within the grounds? Besides, fields of industrial achievement are opened by each investigation, enhancing human welfare, and absorbing our energies. The soldiers of this noble war do not meditate and idealize; their prayer and song are an impulse, not an occupation.

My romancer and scientist goes on to say, "In all this the poet loses nothing. It is fundamental fact that the conquest of mystery leads to greater mystery; the more we know the greater the material for the imagination." This I too believe, and that the poet's province is, and ever must be, the expression of the manner in which revealed truths, and truths as yet unseen but guessed and felt by him, affect the emotions and thus sway man's soul.

<small>The poet's inalienable ground.</small>

Therefore his final ground is still his own, and he well may say, as Whitman chanted thirty years ago: —

> "Space and Time! now I see it is true, what I guessed at.
> . . . .
> I accept Reality, and dare not question it,
> Materialism first and last imbuing.
> . . . .
> Gentlemen, to you the first honors always!
> Your facts are useful, and yet they are not my dwelling,
> I but enter by them to an area of my dwelling.
> Less the reminders of properties told my words,
> And more the reminders they of life untold."

Insight and spiritual feeling will continue to precede discovery and sensation. In their footprints the investigator must advance for his next truth, and at the moment of his advance become one with the poet. In the words of Tyndall on Emerson, "Poetry, with the joy of a bacchanal, takes her grave brother by the hand, and cheers him with immortal laughter." Meanwhile the laws of change, <small>Ebb and flow.</small> fashion, ennui, that breed devotion first to one exercise of man's higher faculty, and anon to another, will direct the public attention alternately to the investigator and to the poet. In lulls or fatigue of discovery, there will be an eager return to the oracles for their interpretation of the omens of the laboratory and ward. The services of the temple are confined no more to the homily and narrative than to song and prayer.

II.

WHAT IS POETRY?

THESE lectures, as I have intimated, are purposely direct of statement, and even elementary. From my point of view this does not of itself imply a lack of respect for the intelligence of the listener. The most advanced star-gazer holds to his mathematics; while, as to poetry, enthusiasts find it easier to build fine sentences than to make clear to others, if to themselves, the nature of that which affects them so inspiringly. I trust that you are willing, in place of the charm of style and the jest and epigram of discourses for entertainment, to accept a search for the very stuff whereof the Muse fashions her transubstantial garments — to discover what plant or moth supplies the sheeny fibre; in what heat, what light, the iridescent fabric is dyed and spun and woven. *A word beforehand.*

It has occurred to me — I think it may not seem amiss to you — that this eager modern time, when the world has turned critic, this curious evening of the century, when the hum of readers and the mists of thought go up from every village; when poetry is both read and written, whether well or ill, more generally than ever before; *The direct and timely question.*

and when clubs are formed for its study and enjoyment, where commentators urban or provincial, masters and mistresses of analytics, devote nights to the elucidation of a single verse or phrase — it has occurred to me that this is an opportune time for the old question, so often received as if it were a jet of cold water upon steam or the stroke of midnight at a masquerade — an apt time to ask ourselves, What, then, *is* poetry, after all? What are the elements beneath its emotion and intellectual delight? Let us have the primer itself. For, if such a primer be not constructible, if it be wholly missing or disdained, you may feel and enjoy a poem, but you will hardly be consistent in your discourse upon it, and this whether you concern yourself with Browning, or Meredith, or Ibsen, — as is now the mode, — or with the masterworks of any period.

Nevertheless, we too must begin our answer to the question, What is poetry? by declaring that the essential *spirit* of poetry is indefinable. It is something which is perceived and felt through a reciprocal faculty shared by human beings in various degrees. The range of these degrees is as wide as that between the boor and the sensitive adept — between the racial Calibans and Prosperos. The poetic spirit is absolute and primal, acknowledged but not reducible, and therefore we postulate it as an axiom of nature and sensation.

The poetic spirit not reducible to terms.

To state this otherwise: it is true that the poetic

essence always has been a force, an energy, both subtile and compulsive; a primal force, like that energy the discovery of whose unities is the grand physical achievement of this century. The shapes which it informs are Protean, and have a seeming elusiveness. Still, even Proteus, as Vergil tells us, is capturable. Force, through its vehicle of light, becomes fixed within the substance of our planet; in the carbon of the fern, the tree, the lump of coal, the diamond. The poetic spirit becomes concrete through utterance, in that poetry which enters literature; that is, in the concrete utterances of age after age. Nothing of this is durably preserved but that which possesses the crystalline gift of receiving and giving out light indefinitely, yet losing naught from its reservoir. Poetry is the diamond of these concretions. It gives out light of its own, but anticipates also the light of after-times, and refracts it with sympathetic splendors. *Its vocal expression may be defined.*

With this uttered poetry, then, we are at present concerned. Whether sung, spoken, or written, it is still the most vital form of human expression. One who essays to analyze its constituents is an explorer undertaking a quest in which many have failed. Doubtless he too may fail, but he sets forth in the simplicity of a good knight who does not fear his fate too much, whether his desert be great or small.

In this mood seeking a definition of that poetic utterance which is or may become of record, — a definition both defensible and *A definition of Poetry in the Concrete.*

inclusive, yet compressed into a single phrase, — I have put together the following statement:

Poetry is rhythmical, imaginative language, expressing the invention, taste, thought, passion, and insight, of the human soul.

<small>1. The imaginative invention and expression are Creative.</small>

First of all, and as a corollary, — a resultant from the factors of imaginative invention and expression, — we infer that poetry is, in common with other art products, a creation, of which the poet is the creator, the maker. *Expression* is the avowed function of all the arts, their excuse for being; out of the need for it, art in the rude and primitive forms has ever sprung. No work of art has real import, none endures, unless the maker has something to say — some thought which he must express imaginatively, whether to the eye in stone or on canvas, or to the ear in music or artistic speech; this thought, the imaginative conception moving him to utterance, being his creative idea — his art-ideal. This simple truth, persistently befogged by the rhetoric of those who do not "see clear and think straight," and who always underrate the strength and beauty of an elementary fact, is the

<small>Metrical sterility.</small>

last to be realized by commonplace mechanicians. They go through the process of making pictures or verses without the slightest mission — really with nothing to say or reveal. They mistake the desire to beget for the begetting power. Their mimes and puppets have everything but souls. Now, the imaginative work of a true artist, conveying his own ideal, is creative because it is the expression, the new embodiment, of his particular nature, the materialization of something which renders him a congener, even a part, of the universal soul — that divinity whose eternal function it is to create. The expressive artist is to this extent indeed fashioned after his Maker. He can even declare, in the words of Beddoes, who used them, however, to reveal his surprising glimpses of evolution: —

> "I have a bit of *Fiat* in my soul,
> And can myself create my little world."

At the same time, the quality of the poet's creation, be it lyrical, narrative, or dramatic, is in a sense that of revelation. He cannot invent forms and methods and symbols out of keeping with what we term the nature of things; such inventions, if possible, would be monstrous, baleful, not to be endured. But he utters, reveals, and interprets what he sees with that inward vision, that second sight, the prophetic gift of certain personages, — that which I mean by "insight," and through which the poet is thought to be inspired. This vision penetrates what Plato conceived to be the quintessence of nature, what Wordsworth, in his very highest mood, declares that we perceive only when

2. The poet a revealer through Insight.

> "we are laid asleep
> In body, and become a living soul:
> While with an eye made quiet by the power
> Of harmony, and the deep power of joy,
> We see into the life of things."

The creative insight, according to its degree, is
allied with, if not the source of, the mysterious endowment named genius, which humdrum intellects have sought to disallow, claiming that it lies chiefly in one of its frequent attributes, — industry, — but which the wisdom of generations has indubitably recognized. The antique and idealistic notion of this gift is given in "Ion": "A poet ... is unable to compose poetry until he becomes inspired and is out of his sober senses, and his imagination is no longer under his control; for he does not compose by art, but through a divine power." The modern and scientific rendering is that of the exact investigator, Hartmann, who traces this power of genius to its inmost cell, and classifies it as the spontaneous, involuntary force of the untrammelled soul, — in precise terms, "the activity and efflux of the Intellect freed from the domination of the Conscious Will." Whichever statement you accept, — and I see no reason why the two are not perfectly concordant, — here is the apparently superhuman gift which drew from Sophocles that cry of wonder, "Æschylus does what is right without knowing it."

Genius, its recognition and its scientific vindication.

As an outcome of genius producing the semblance of what its insight discovers, poetry aims to convey beauty and truth in their absolute simplicity of kind, but limitless variety of guise and adaptation. The poet's vision of these is shared to some extent by all of us, else

3. Poetry as an expression of the beautiful,

his appeal would not be universal. But to *his* inborn taste and wisdom is given the power of coadequate expression. Taste has been vilely mistaken for a sentiment, and disgust with its abuse may have incited the Wordsworthians and others to disqualify it. They limited their own range by so doing. The world forgives most sins more readily than those against beauty. There was something ridiculous, if heroic, in the supercilious attitude of our transcendentalists, not only putting themselves against the laity, but opposing the whole body of their fellow seers and artists, whose solace for all labors ever has been the favor of their beloved mistress Beauty,— the inspirer of creative taste.

The truth is that taste, however responsive to cultivation, is inborn,— as spontaneous as insight, and, indeed, with an insight of its own. Schlegel's alertness with respect to the æsthetic moved him to define even genius as "the almost unconscious choice of the highest degree of excellence, and, consequently," he added, "it is taste in its highest activity." Profound thinkers, lofty and unselfish natures, may flourish without taste: if so, they miss a sense, nor only one that is physical, — something else is lacking, if the body be the symbol of the soul. I would not go so far as to say of one born, for instance, without ear for melody, that there will be "no music in his soul" when that is disembodied. It is finer to believe that

through creative Taste.

> "whilst this muddy vesture of decay
> Doth grossly close it in"

such a one cannot hear it; that

> "The soul, with nobler resolutions deck'd,
> The body stooping, does herself erect."

But taste, whether in or out of the body, is a faculty for want of which many ambitious thinkers have in the end failed as poets. It is a sense, however, the functions of which are very readily assumed and mechanically imitated. At periods when what are called false and artificial standards have prevailed, as, in French and English letters from 1675 to 1790, the word "taste" has been on every one's lips, and the true discernment of beauty has been supposed to be supreme, when in fact merely the crown and sceptre of taste have been set up and its mantle stuffed with straw. At this very time art is suffering everywhere from an immense variety of standards and models, and our taste, in spite of the diverse and soulless yet attractive productions of the studio and the closet, is that of an interregnum.

Taste often wanting or assumed.

Assuming that the artist's conceptions are spontaneous and imaginative, their working out brings into play the conscious intellect. He gives us thought, building up masterpieces from the germinal hint or motive: his wisdom is of so pure a type that through it the poet and the philosopher, in their ultimate and possible development, seem united. It is the exclusive presentation of thought and truth that makes poetry didactical, and hence untrue in the artistic sense.

4. Poetry as an expression of intellectual Thought.

For taste has been finely declared to be "the artistic ethics of the soul," and it is only through a just balance of all the elements in question that poetry rises above ordinary and universal human speech and becomes a veritable art.

Under the conditions of these reciprocal elements, the poet's nature, "all touch, all eye, all ear," exalted to a creative pitch, becomes *emotional*. Feeling is the excitant of genuine poetry. The Miltonic canon, requiring the sensuous beauty which taste alone insures, demands, last of all, as if laying stress upon its indispensability, that poetry should be passionate. It is the impassioned spirit that awakes the imagination, whose taste becomes alert, that hears whisperings which others do not hear, — which it does not itself hear in calmer periods, — that breaks into lyric fervor and melody, and that arouses kindred spirits with recital of its brave imaginings. Feeling of any kind is the touch upon the poet's electric keyboard; the *passio vera* of his more intense moods furnishes the impulse and the power for effective speech. His emotion instinctively acquires the tone and diction fitted to its best expression. Even the passion of a hateful nature is not without a certain distinction. Flame is magnificent, though it feed upon the homes of men.

Right here we stop to consider that thus far our discussion of the poetic elements applies with almost

<small>5. Emotion. The poet must be impassioned.</small>

equal significance to all the fine arts; each of them, in fact, being a means of expressing the taste, thought, passion, imagination, and insight, of its devotee. The generic principles of one are those of all. Analysis of one is to this extent that of art as art: a remark illustrated by the talk of every noteworthy virtuoso, from Angelo to Reynolds and Ruskin and Taine. Reflect for an instant upon the simultaneous appearance of a certain phase, such as Preraphaelitism, in the plastic, structural, and decorative arts, in imaginative literature; and on the stage itself, and you see that the Muses are indeed sisters, and have the same food and garments, — often the same diseases. But take for granted the "consensus of the arts." What is it, then, that differentiates them? Nothing so much as their respective vehicles of expression. The key-stone of our definition is the statement that poetry, in the concrete and as under consideration, is *language*. Words are its specific implements and substance. And art must be distinguished, whatsoever its spirit, by its concrete form. A picture of the mind is not a painting. There is a statue in every stone; but what matters it, if only the brooding sculptor sees it? A cataract, a sunset, a triumph, a poetic atmosphere, or mood, or effect, — none of these is a poem. When Emerson and Miss Fuller went together to see Fanny Elssler dance, and the philosopher whispered to the sibyl, "Margaret, this is poetry!" and the sibyl re-

Marginal notes: But the foregoing elements pertain to all the arts. — 6. Poetry, then, is an art of Speech.

joined, "Waldo, it is religion!" they both, I take it, would have confessed with Hosea that they had used similitudes. We are now considering the palpable results of inspiration. Poetry houses itself in *words*, sung, spoken, or inscribed, though there is a fine discrimination in the opening sentence of Ben Jonson's Grammar, which declares of language that "the writing [of it] is but an accident."

Language is colloquial and declarative in our ordinary speech, and on its legs for common use and movement. Only when it takes wings does it become poetry. As the poet, touched by emotion, rises to enthusiasm and imaginative power or skill, his speech grows *rhythmic*, and thus puts on the attribute that distinguishes it from every other mode of artistic expression — the guild-mark which, rightly considered, establishes the nature of the thing itself. At this date there is small need to descant upon the universality of rhythm in all relations of force and matter, nor upon its inherent consonance with the lightest, the profoundest, sensations of the living soul. Let us accept the wisdom of our speculative age, which scrutinizes all phenomena and reaches the scientific bases of experience, and, looking from nadir to zenith, acknowledges a psychological impulse behind every physical function. The earliest observers saw that life was rhythmical, that man and brute are the subjects of recurrent touch, sensation, order, and are alike responsive to measured sound, the form of

Its characteristic language always Rhythmical.

rhythm most obvious and recognizable; that music, for instance, affects the most diverse animate genera, from the voiceless insect and serpent to the bird with its semi-vocal melody, and the man whom it incites to speech and song. The ancients no less comprehended the rhythm of air and water, the multitudinous harmonies, complex and blended, of ocean surges and wind-swept pines. But our new empiricism, following where intuition leads the way, comprehends the function of *vibrations:* it perceives that every movement of matter, seized upon by universal force, is *vibratory;* that vibrations, and nothing else, convey through the body the look and voice of nature to the soul; that thus alone can one incarnate individuality address its fellow; that, to use old Bunyan's imagery, these vibrations knock at the ear-gate, and are visible to the eye-gate, and are sentient at the gates of touch of the living temple. The word describing their action is in evidence: they "thrill" the body, they thrill the soul, both of which respond with subjective, interblending vibrations, according to the keys, the wave-lengths, of their excitants.

<small>The soul responsive to Vibrations.</small>

Thus it is absolutely true that what Buxton Forman calls "idealized language," that is, speech which is imaginative and rhythmical, goes with emotional thought; and that words exert a mysterious and potent influence, thus chosen and assorted, beyond their normal meanings. Equally true it is that natural poets in sensi-

<small>Every true poet is born with the gift of Rhythm.</small>

tive moods have this gift of choice and rhythmic assortment, just as a singer is born with voice and ear, or a painter with a knack of drawing likenesses before he can read or write. It is not too much to say that if not born with this endowment he is not a poet: a poetic nature, if you choose,— indeed, often more good, pure, intellectual, even more sensitive, than another with the "gift,"— and, again, one who in time by practice may excel in rhythmical mechanism him that has the gift but slights it; nevertheless, over and over again, not a born poet, not of the royal breed that by warrant roam the sacred groves. I lay stress upon this, because, in an age of economics and physics and prose fiction, the fashion is to slight the special distinction of poetry and to deprecate its supremacy by divine right, and to do this as our democracy reduces kingcraft — through extending its legitimate range. You cannot force artists, architects, musicians, to submit to such a process, for material dividing lines are too obvious. Otherwise, some would undoubtedly make the attempt. But poetic vibrations are impalpable to the carnal touch, and unseen by the bodily eye, so that every realist, according to his kind, either discredits them or lays claim to them. All the same, nothing ever has outrivaled or ever will outrival, as a declaration of the specific quality of poetry, the assertion that its makers do

"feed on thoughts, that *voluntary* move
Harmonious numbers;"

and the minstrel poet, of my acceptation, "lisped in numbers" as an infant — and well does the hackneyed verse reiterate, "for the numbers came."

Aside from the vibratory mission of rhythm, its <small>Rhythmical factors and minor aids.</small> little staff of adjuvants, by the very discipline and limitations which they impose, take poetry out of the place of common speech, and make it an art which lifts the hearer to its own unusual key. Schiller writes to Goethe that "rhythm, in a dramatic work, treats all characters and all situations according to one law. . . . In this manner it forms the atmosphere for the poetic creation. The more material part is left out, for only what is spiritual can be borne by this thin element." In real, that is, spontaneous minstrelsy, the fittest assonance, consonance, time, even rhyme, — if rhyme be invoked, and rhyme has been aptly called "both a memory and a hope," — come of themselves with the imaginative thought. The soul may conceive unconsciously, and, as I believe in spite of certain metaphysicians, without the use of language; but when the wire is put up, the true and only words — just so far as the conception is true and clear and the minstrel's gift coequal — are flashed along it. Such is the test of genuineness, the underlying principle being that the masterful words of all poetic tongues are for the most part in both their open and consonantal sounds related to their meanings, so that with the inarticulate rhythm of impassioned thought we have a correspondent verbal rhythm for its vehicle. The whole range of poetry which is vital, from the Hebrew psalms and prophecies, in their original text and in our great English version, to the Georgian lyrics and romances and the Victorian idyls, confirms the statement of Mill that "the deeper the feeling, the more characteristic and decided the rhythm." The rapture of the poet governs the tone and accent of his

> "high and passionate thoughts
> To their own music chanted."

Whoever, then, chooses to exempt poetry from this affinity with rhythm is not consider- *The essential* ing the subject-matter of these discourses. *differentiation.* Not that I would magnify its office, or lessen the claims of other forms of imaginative and emotional expression. "The glory of the celestial is one, and the glory of the terrestrial is another.... There is one glory of the sun, and another glory of the moon." Nor do I ask you, with the Scripture, to set one above the other: count them of equal rank, if you like, — as in truth they seem to be in a time which has produced not only "In Memoriam," "Pippa Passes," "The Problem," but also "A Tale of Two Cities," "Henry Esmond," "The Scarlet Letter,"— but count them as *different*. Of one thing I am assured, that every recognized poet will claim the vitality of this difference — a professional claim, without doubt, but not as though made by a lawyer or a divine, since their professions are more arbitrary and acquired. I confess that natural aptitude justifies in a measure the expressions "a born lawyer," "a born doctor," etc.; still, more of what we call professional skill is obtained by training than by derivation. The reverse of this is true of minstrelsy, and thus it chances that for a thousand excellent lawyers you shall not discover one superior poet.

It is not essential now, when the trick of making *Rhyme, etc.* clever verse is practised, like all the minor technics of decoration, music, and so on, by many more or less cultured persons with a talent for mimicry, to discuss historic forms of measure, and to show why rhythm is not confined to any formal measures rhymed or unrhymed. Yet even rhyme, in our tongue, has advantages apart from its sound, when so affluent and strong a workman as Browning uses it in some of his most extended poems as a brake on the whirl and rush of an over-productive genius. All the varied potencies of rhythm, — its trinity of time-beat, consonance, and assonance, its repetends and refrains and accidental wandering melodies and surprises, — are the vibrations of the poetic fervor made manifest, and the poet's conveyance of it to his listeners.

Now, we have seen that the term poetry was long *Imaginative prose fiction:* applied to all imaginative literature. I recognize the fact that the portion of it which was only germinal with the ancients, but is the chief characteristic of our modern age, the prose tale or romance, — that this, our prose fiction, is equally a part of the feigned history celebrated by

Plato and Bacon and Sidney, of the thing creatively invented rather than of things debated or recorded. It is often imbued with the true spirit of poesy, and is almost always more original in plot, narrative, structure, than its sister art. It well may supply the topic for a series of discourses. Among the brilliant romancers and novelists are not a few who, were not fiction the dominant mode of our time, would possibly have wreaked their thoughts upon expression in rhythmical form. But to see how distinct a thing it is, and also to illustrate my belief that a dramatic poet may as well not originate his own narrative or plot, read a story of Boccaccio or a chronicle by Holinshed, and then the play of Shakespeare's moulded upon it. The masterly novelist, the better to control his plot and to reflect life as it is, keeps his personal emotion within such command that it fails to become rhythmical. Where it gets the better of him, and he breaks into blank verse or singsong, his work is infallibly weakened; it may catch the vulgar ear, but is distinctly the less enduring. Who now can abide the tricky metrical flow of certain sentimental passages in Dickens? And Dickens, by the way,— nature's own child and marvellous, as in truth he was,— occasionally set himself to write poetic verse, but he knew no trick of it, and could acquire none. His lyrics were mostly commonplace. This was to be expected, for a real poet usually writes good prose, and rarely rhythmical prose as prose, though

how it is distinguished from the poetry under consideration.

he may elect, with Macpherson, Blake, Tourgénieff, Emerson, and Whitman, to cast his poetry in rhythmical prose form. Thackeray, who was a charming poet, of a light but distinct quality above which he was too genuine to venture, put no metrics into his novels. See how definite the line between the prose and the verse of Milton, Goethe, Landor, Coleridge, Byron. Of Emerson I have said elsewhere that his prose was poetry, and his poetry light and air. There is a class of writers, of much account in their day, whose native or purposed confusion between rhythmical and true prose attracts by its glamour, and whom their own generation, at least, can ill spare. Of such was Richter, and such in a measure have been De Quincey, Wilson, Carlyle, and even Ruskin, each after his kind. The strong personality of a writer forces its way. But it is to be noted that these after a time fall into distrust, as if the lasting element of true art had somehow escaped them. Certain latter-day lights well might take a lesson from the past. These illuminati leave firm ground, but do they rise to the upper air? There is something eerie and unsubstantial about them as they flit in a moonlit limbo between earth and sky. Howbeit, they are what they are, and may safely plead that it is more to be what they can be than not to be at all. The difference betwixt poetical prose and the prose of a poet is exemplified by Mark Pattison's citation of the two at their best — the prose of Jeremy

The prose of poets, vs. "poetical" prose.

Taylor and that of Milton, the former "loaded with imagery on the outside," but the latter "colored by imagination from within."

In short, although throughout our survey, and especially in the Orient, the most imaginative poetry often chants itself in rhythmic prose, the less rhythm there is in the prose of an essayist or novelist the better, even though it characterizes an interlude. As a drop of prosaic feeling is said to precipitate a whole poem, so a drop of sentimental rhythm will bring a limpid tale or essay to cloudy effervescence. As for eloquence, which also was classed with poetry by our ancestors, and which is subjective and passionate, I do not say that it may not rise by borrowing wings; but in a poem the force of eloquence pure and simple cannot be prolonged without lessening ideality and the subtlest quality of all, — suggestiveness, — and rhetoric is as false a note as didacticism in the poet's fantasia. *Eloquence, rhetoric, etc.*

It is worth while to observe, in passing, that there never was a time before our own in literary history when more apparent successes, more curious and entertaining works, were achieved by determined and sincere aspirants who enter, not through original bent, but under gradual training and "of malice aforethought," fields to which they are not born inheritors, — the joint domains of poetry and prose fiction. Their output deceives even the critic, because it does serve a *Modern cleverness and training.*

purpose, until he reflects that none of it is really a force,—really something new, originative, enduring. Such a force was that of Fielding, of Byron, of Scott, of Keats, of Wordsworth, of Browning; and many lesser but fresh and natural poets and novelists are forces in their several degrees. What they produce, from its individual, often revolutionary, quality, is an actual addition to literature. But we see natural critics and moralists, persons of learning, of high cultivation in the focal centres of literary activity, who develop what *is* inborn with them —an exquisite gift of appreciation, and in time a stalwart purpose to rival the poets and novelists on their own ground. This they undertake at that mature age when the taste and judgment are fully ripe, and after admirable service as scholars, essayists, and the like. Now, there scarcely is an instance, in the past, of a notable poet or romancer who did not begin, however late, by producing poetry or fiction, however crude, and this whether or not he afterward made excursions into the fields of analysis or history or æsthetics. Mr. Howells is a living illustration of this natural process. He began as a poet, and then, after excursions into several literary fields that displayed his humor, taste, and picturesqueness, he caught the temper of his period as a novelist, and helped to lead it. The cleverness and occasional "hits" of various self-elected poets and tale-writers are, however, noteworthy, even bewildering. At this mo-

<small>Nature's process.</small>

ment many who command public attention and what is called the professional market have previously demonstrated that their natural bent was that of didactic and analytic, rather than of emotional and creative, writers. Their success has been a triumph of culture, intellect, and will power. These instances, as I have said of an eminent poet and essayist now no more, almost falsify the adage that a poet is born, not made. Still, we bear in mind that precisely analogous conditions obtain in the cognate artistic professions, — in painting, music, architecture. The poets and novelists by cultivation, despite their apparent vogue in the most extended literary market the world has ever seen, and ambitious as their work may be, lack, in my opinion, the one thing needed to create a permanent force in the arts, and that is the predestined call by nature and certain particles of her "sacred fire." *Cp. "Victorian Poets": pp. 91, 442.*

We need not enter the poet's workshop and analyze the physics and philosophy controlling the strings of his lyre. That a philosophical law underlies each cadence, every structural arrangement, should be known in this very spot,[1] if anywhere, where not alone the metrics and phonetics, and what has been called the rationale, of verse, but therewithal the spirit of the poetry of the East, of our classical antiquity, of the Romance tongues, of the Norse, and of our own composite era, are in *"The Science of Verse."*

[1] Johns Hopkins University.

the air, one may say, and are debated with a learning and enthusiasm for which a few of us, in my own academic days, hungered in vain. Here, too, it was that the most analytic treatise ever conceived, upon the technics of rhythmical effect, was written by *Lanier.* your own poet, Lanier, for whom the sister-spirits of Music and Poesy contended with a rivalry as strong as that between "twin daughters of one race," both loving, and both worshipped by, one whom death too soon removed while he strove to perfect their reconciliation. Though poetry must come by the first intention, if at all, and inspiration laughs at technical processes, even the unlettered minstrel conforms to law, as little conscious of it as some vireo in the bush is conscious of the score by which a Burroughs or an Olive Miller transfers the songster's tirra-lirra to the written page. The point remains that poetry is ideal expression through *Poetry, above all, is utterance.* words, and that words are not poetry unless they reach a stress that is *rhythmical*. Painting is a mode of expression, being visible color and shadow distributed upon a material surface; the language of poetry is another mode, because it is *articulate* thought and feeling. Sidney pointed merely to the fact that rhythm is not confined to verse, when he spoke of "apparelled verse" as "an ornament, and no cause to poetry, since there have been many most excellent poets that never have versified"; and he added that "now swarm many versifiers that need never answer to the name of poet." Wordsworth's familiar recognition of "the poets that ne'er have penned their inspiration" was a just surmise; but such a poet is one *in posse*, assuredly not *in esse*, not a maker. Swinburne traverses the passage with a bit of common sense — "There is no such thing as a dumb poet or a handless painter. The essence of an artist is that he should be articulate."

Submitting these views with respect to a scientific definition of poetry, I ask your attention to a brief consideration of its bounds and liberties, as compared with those of music and the respective arts of design. <small>Comparative review of the Arts.</small>

The specific province, by limitation, of Sculpture, the art consecrate to the antique precision of repose, is to express ideals of form *arrested as to movement and time*. Its beauteous or heroic attitudes are caught at the one fit moment, and forever transfixed in rigid stone or wood or metal. Painting has an additional limitation; it gives only the *similitude* of form in all its dimensions, and only from one point of a beholder's view. To offset this, the range of the painter is marvellously broadened by the truth of perspective, the magic and vital potency of color, the tremulous life of atmosphere, and the infinite gradations and contrasts of light and shade. The mystical warmth and force of the Christian humanities are radiant in this enrapturing art. Yet its office is to capture the <small>The respective powers and limitations of Sculpture, Painting,</small>

one ideal moment, the lifelong desire of Faust, and to force it to obey the mandate:—

"Ah, still delay—thou art so fair!"

Such are the arts addressed to the eye alone, both of them lending their service to the earliest, the latest, the most various, of all material constructions — Architecture, whose pediments and roofs and walls originate in our bodily necessities, whose pinnacles typify our worship and aspiration, and which so soon becomes the beneficiary and the incasement of its decorative allies. None of the three can directly express time or movement, but there is practically no limit to their voiceless representation of space and multitude.

<small>Architecture.</small>

But movement in time is a special function of Music, that heavenly maid, never so young as now, and still the sovereign of the passions, reaching and rousing the soul through sound-vibrations perpetually changing as they flow. To this it adds the sympathetic force of harmonic counterpoint. Its range, then, is freer than that of the plastic and structural arts, by this element of progressive change. Under its spell, thrilling with the sensations which it can excite, and which really are immanent in our own natures, considering moreover the superb mathematics of its harmony, and again that it has been the last in development of all these arts, we question whether it is not only superior to them

<small>Music.</small>

<small>The composer's sublime freedom, through progressive change.</small>

but even to that one to which these lectures are devoted. All feel, at least, the force of Poe's avowal that music and poetry at their highest must go together, because "in music the soul most nearly attains the great end for which it struggles — supernal beauty." And so old John Davies, in praise of music, —

> " The motion which the ninefold sacred quire
> Of angels make: the bliss of all the blest,
> Which (next the Highest) most fills the highest desire."

Schopenhauer thought that the musician, because there is no sound in nature fit to give him more than a suggestion for a model, "approaches the original sources of existence more closely than all other artists, nay, even than Nature herself." Herbert Spencer has suggested that music may take rank as the highest of the fine arts, as the chief medium of sympathy, enabling us to partake the feeling which excites it, and "as an aid to the achievement of that higher happiness which it indistinctly shadows forth." And in truth, if the intercourse of a higher existence is to be effected through sound-vibrations rather than through the swifter light-waves, or by means of aught save the absolute celestial insight, one may fondly conceive music to be the language of the earth-freed, as of those imagined seraphim with whom feeling *is* "deeper than all thought."

Consider, on the other hand, how feeling governs the simple child, "that lightly draws its breath," while thought begins its office as the child grows in strength and knowledge, and it is a fair inference that thought is the higher attribute, and that the suggestion of emotion by music is a less vital art than that of intellectual speech. The dumb brutes partake far more of man's emotion than of his mental intelligence. Neither is music — despite our latter-day theorists who defy the argument of Lessing's Laocoön and would make one art usurp the province of another, and despite its power as an indirect incentive to thought by rousing the emotions — a vehicle for the conveyance of precise and varied ideas. The clearer the idea, the more exact the language which utters and interprets it. This, then, is the obvious limitation of music: it can traverse a range of feeling that may govern the tone of the hearer's contemplations, it can "fill all the stops of life with tuneful breath" and prolong their harmonic intervals indefinitely, but the domain of absolute *thought*, while richer and more imperial for its excitation, is not mastered by it. Of that realm music can make no exact chart.

<small>But intellectual Speech is supreme.</small>

Thus far, we have no art without its special office, and none that is not wanting in some capacity displayed by one or more of the rest. Each goes upon its appointed way. Now comes poetry, — rhythmical, creative language, compact of beauty, imagination, passion, truth, — in no wise related, like the plastic arts, to material substance; less able than its associate, music, to move the soul

<small>Limitations of the poet.</small>

with those dying falls of sound that increase and lessen thought and the power to harbor it; almost a voiceless spirit of invention, working without hands, yet the more subtile, potent, inclusive, for this evasive ideality, and for creations that are impalpable except through the arbitrary and non-essential symbols by which it now addresses itself to the educated eye.

Permit me to select, almost at random, from Keats and Tennyson, ready illustrations of the bounds and capabilities of the various arts — passages necessarily familiar, since they *are* from Keats and Tennyson, but chosen from those masters because, of all English poets since Spenser, they are most given to picture-making, to the craft that is, as we say, artistic, picturesque. A stanza from the "Ode on a Grecian Urn" describes, and rivals in verse, the ravishing power of a bit of sculpture to perpetuate arrested form and attitude — yes, even the suggestion of arrested music : — *[How far he can illustrate and imitate sculpture.]*

> "Heard melodies are sweet, but those unheard
>   Are sweeter; therefore, ye soft pipes, play on —
> Not to the sensual ear, but, more endear'd,
>   Pipe to the spirit ditties of no tone.
> Fair youth, beneath the trees, thou canst not leave
>   Thy song, nor ever can those trees be bare;
> Bold lover, never, never canst thou kiss,
>   Though winning near the goal; yet, do not grieve —
> She cannot fade, though thou hast not thy bliss;
>   Forever wilt thou love, and she be fair."

These undying lines not only define by words the

power and limits of the sculptor, but are almost a matchless example of the farthest encroachment poetry can make upon sculpture's own province.[1] What it cannot do is to combine the details of the carving so as to produce them to the mind, as sculpture does to the eye, at a single instant of time. It lingers exquisitely upon each in succession. Progressive time is required for its inclusion of the effects of a Grecian frieze or scroll. Now, take from Tennyson's lovely but lighter poem, "The Day-Dream," — a lyrical idyl at the acme of melodious and fanciful picture-making, — a stanza which seems to match with a certain roundness and color the transfixing effect of the painter's handiwork. It portrays a group entranced by the spell that has doomed to a hundred years of abeyance and motionlessness the life of the king's palace and the Sleeping Beauty. In the poems of Keats and Tennyson, as I say, artists find their sculptures and paintings already designed for them, so that these poets are the easiest of all to illustrate with some measure of adequacy. The theme of the following lines, rendered by a painter, would show the whole group and scene at a flash of the eye; poetry cannot do this, yet, aided

His picture-making:

its liberties and bounds.

[1] Since the first appearance of this lecture I have seen a finely penetrative essay by Mr. J. W. Comyns Carr (*The New Quarterly Magazine*, October, 1875), in which this same Ode is quoted to illustrate the ideal calm sought for by "The Artistic Spirit in Modern Poetry." As no better example can be found, in conveyance of the poetic and the plastic methods respectively, I do not hesitate to retain it.

by its moving panorama, the listener has painted all in his mind when the last word is uttered : —

> "More like a picture seemeth all
> Than those old portraits of old kings,
> That watch the sleepers from the wall.
>
> "Here sits the butler with a flask
> Between his knees, half-drain'd ; and there
> The wrinkled steward at his task,
> The maid-of-honor blooming fair;
> The page has caught her hand in his :
> Her lips are sever'd as to speak :
> His own are pouted to a kiss :
> The blush is fix'd upon her cheek."

It is to be noted, as we read, that Tennyson's personages, and those of Keats as well, are mostly conventional figures, as characterless as those on a piece of tapestry. The genius of neither poet is preferably dramatic : they do not get at individuality by dramatic insight like Shakespeare, nor by monodramatic soliloquy and analysis, like the strenuous Browning. Their dramas are for the most part masques containing *eidullia* (little pictures); though who can doubt that Keats, had he lived, would have developed the highest dramatic power? Remember what the less sensuous, more lyrical Shelley achieved in "The Cenci," when only four years beyond the age at which Keats imagined his " gold Hyperion, love-lorn Porphyro." But, to resume, see what poetry, in addition to the foregoing counterfeit of the painter's ocular presentment, can bring about in its

<small>Artist-poets.</small>

<small>The poet infuses Life by his command of vocal movement.</small>

own field through its faculty of *movement in time* — a power entirely wanting to the arts which it has just mimicked. Note how it breaks the spell of transfixed attitude, of breathless color and suspended action; how it lets loose the torrents of Life at the instant of the "fated fairy prince's" experimental kiss:—

> "A touch, a kiss! the charm was snapt.
>   There rose a noise of striking clocks,
> And feet that ran, and doors that clapt,
>   And barking dogs, and crowing cocks;
> A fuller light illumined all,
>   A breeze thro' all the garden swept,
> A sudden hubbub shook the hall,
>   And sixty feet the fountain leapt.
> . . . .
> The maid and page renew'd their strife,
>   The palace bang'd, and buzz'd, and clackt,
> And all the long-pent stream of life
>   Dash'd downward in a cataract."

That is the stream which the painter has no art to undam. Only by a succession of pictures can he suggest its motion or follow the romance to its sequel; and that he can do even this with some fitness in the case of a Tennysonian ballad is because the laureate, as we see, counterfeits the painter's own method more artistically than other idyllists of rank in our time. If art is the fit and beautiful conformation of matter infused with the spirit of man, it must indeed have life. The most nimble, ardent, varied transfer of the vital spirit is by means of language, and of all language that of

He is again supreme.

the poet is the most alive and expressive. Observe, again, that in what are called art circles — Arcadian groups of those devoted to art and letters — the imaginative writers are apt to interest themselves far more with respect to the plastic arts than the sculptors and painters with respect to poetry and romance; and well they may, since the poet enriches his work by using all artistic effects, while nothing is more dangerous to a painter, for example, than that he should give his picture a literary cast, as the phrase is, and make it too closely tell a story or rehearse a poem. This of itself tends to confirm Lessing's apothegm that "the poet is as far beyond the painter as life is better than a picture."

THE conquests of poetry, in fine, are those of pure intelligence, and of emotion that is unfettered. Like the higher mathematics, it is not dependent on diagrams, for the mind to which it appeals is a responsive draughtsman of lines finer and more complex than any known to brush or graver. It creates no beauty of form beyond the accidental symbols grouped in script and print, none of light and color, while the ear is less touched by it than by the melodies or harmonies of music; for its melody is that of flexible speech, and it knows not counterpoint, but must resort to the value of successive strains. Yet we

Final analysis and summary of the chief of arts.

say that it has form and outline of its own, an architecture of its own, its own warmth and color, and, like music, life, and withal no little of music's vocal charm, in that through words it idealizes these "sweet influences," and is chartered to convey them all to the inward sight, the spiritual hearing, of the citadeled soul, with so apt suggestion that the poet's fellow-mortals, of like passions and perceptions with himself, see and hear and feel with much of his distinct individuality. Its vibrations excite the reflex action that creates in the mind of the receiver a vision corresponding to the imagination of the poet. Here is its specific eminence: it enables common mortals to think as the poet thinks, to use his wings, move through space and time, and out of space and time, untrammelled as the soul itself; it can feel, moan, weep, laugh, be eloquent, love, hate, aspire, for all — and with its maker; can reflect, and know, and ever seek for knowledge; can portray all times and seasons, and describe, express, interpret, the hiddenmost nature of man. Through poetry soul addresses soul without hindrance, by the direct medium of speech. Words are its atmosphere and very being: language, which raises man above the speechless intelligences; which, with resources of pitch, cadence, time, tone, and universal rhythm, is in a sense a more advanced and complex music than music itself — that idealized language which, as it ever has been the earliest form of emotional expression, appears almost a gift captured in man's infancy from some "imperial palace whence he came." To the true poet, then, we say, like the bard to Israfel : —

> " The ecstasies above
> With thy burning measures suit —
> Thy grief, thy joy, thy hate, thy love,
> With the fervor of thy lute —
> Well may the stars be mute."

III.

CREATION AND SELF-EXPRESSION.

THE difficulty that confronts one who enters upon a general discussion of poetry is its universal range. The portals of his observatory tower before him, flashing yet frowning, and inscribed with great names of all the ages. Mount its stairway, and a chart of the field disclosed is indeed like that of the firmament. In what direction shall we first turn? To the infinite dome at large, or toward some particular star or group? We think of inspiration, and a Hebrew seer glows in the prophetic East; of gnomic wisdom and thought, and many fixed white stars shine tranquilly along the equinox, from Lucretius to Emerson; of tragedy and comedy, the dramatic coil and mystery of life, and group after group invites the lens,—for us, most of all, that English constellation blazing since "the spacious times of great Elizabeth"; of beauty, and the long train of poetic artists, with Keats like his own new planet among them, swims into our ken. Asia is somewhere beyond the horizon, and in view are countless minor lights,—the folk-singers and minstrels of many lands and generations.

The radiant dome.

The future lecturer will have the satisfaction of

giving his attention to a single master or school, — to the Greek dramatists, to Dante, or Milton, or Goethe; more than one will expend his resources upon the mimic world of Shakespeare, yet leave as much for his successors to accomplish as there was before. Their privilege I do not assume; since these initiatory discourses have to do with the elements of which poetry is all compact, and with the spirit in fealty to which its orbs shine and have their being and rehearse the burden of their radiant progress: —

> "Beneath this starry arch
> Nought resteth or is still:
> But all things hold their march,
> As if by one great will:
> Moves one, move all: hark to the footfall!
> On, on, forever!"

Two main divisions.

Still, I wish in some way to review this progress of poesy. Essaying then, for the little that can be done, to look first at the broad characteristics of the field, we see that there are, at all events, two streams into which its vast galaxy is divided, — though they intersect each other again and again, and in modern times seem almost blended. These do not relate to the technical classification of poetry: to its partition by the ancients into the epic, dramatic, lyric, and the idyllic, — unto which we have added the reflective, and have merged them all in the composite structures of modern art. Time has shown that we cannot overrate the method of those

A passing reference to established orders.

intuitive pagans. No one cares for Wordsworth's division of his own verse into poems of imagination, of fancy, and the like, the truth being that they all, with the exception of a few spontaneous lyrics, are poems of reflection, often glorified by the imagination, sometimes lightened by fancy, but of whose predominant spirit their author was apparently the least successful judge. The Greeks felt that the spirit shapes the form of art, and therefore is revealed by it. Assume, then, the fitness of poetic orders, styles, and measures; that these are known to you and me, and thus we may leave dactyls and choriambs to the metrical anatomists, and rhymes to the Walkers and Barnums. Passing to the more essential divisions of expression, you will find their types are defined by the amount of personality which they respectively hold in solution: that poetry is differentiated by the Me and the Not Me, — by the poet's self-consciousness, or by the representation of life and thought apart from his own individuality. *But poetry is either impersonal or self-expressive.*

That which is impersonal, and so very great at its best, appears the more creative as being a statement of things discerned by free and absolute vision. The other order is so affected by relations with the maker's traits and tastes that it betokens a relative and conditioned imagination; and is thus by far the larger division, since in most periods it is inevitable that the chief impulse to song should be a conscious or unconscious longing for personal expression.

The gift of unconditioned vision has been vouchsafed both to the primitive world, and to
races at their height of action and invention. The objective masterpieces of poetry consist,
first, of those whose origin is obscure, and which
are so naturally inwrought with history and popular
traits that they seem growths rather than works of
art. Such are the Indian epics, the Northern sagas,
the early ballads of all nations, and of course the
Homeric poems of Greece. These are the lusty
product of the youth of mankind, the
song and story that come when life is un-
jaded, faith unsophisticated, and human nature still
in voice with universal Pan. The less spontaneous
but equally vital types are the fruit of later and constructive periods,—"golden" ages, the masterpieces
of which are composed with artistic design and still
unwearied genius. Whether epic or dramatic, and
whether traditional or the product of schools and
nations in their prime, the significance of objective
poetry lies in its presentment of the world outside,
and not of the microcosm within the poet's self.
His ideal mood is that of the Chinese sage from
whose wisdom, now twenty-six centuries old, the
artist John La Farge, himself imbued with the
spirit of the "most eastern East," has cited for me
these phrases: "I am become as a quiet water, or a
mirror reflecting what may be. It keeps nothing, it
refuses nothing. What it reflects is there, but I do
not keep it: it is not I." And again: "One should

[sidenotes: The Creative and Impersonal. Juventus mundi.]

be as a vacuum, so to be filled by the universe. Then the universe will fill me, and pour out again." Which dark saying I interpret here as an emblem of the receptivity of the artist to life at large. This it is his function to give out again, illumined, but unadulterate. The story is told, the song chanted, the drama constructed, with the simplest of understandings between audience and maker: as between children at their play, artisans at their handicraft, recounter and hearers around the desert fire. Every literature has more or less of this free, absolute poetry. But only in the drama, and at distinctively imaginative periods, have poets of the Christian era been quite objective; not even there and then, without in most cases having "unlocked" their hearts by expression of personal feeling. This process — exemplified in the sonnets of Shakespeare, and in the minor works of Dante, Tasso, Cervantes, Calderon, Camoëns — rarely suggested itself to the antique poets, whose verses were composed for the immediate verdict of audiences great or small, and in the Attic period distinctly as works of art, necessarily universal, and not introspective. Nor would much self-intrusion then have been tolerated. Imagine the Homeric laughter of an Athenian conclave, every man of them with something of Aristophanes in him, at being summoned to listen to the sonnetary sorrows of a blighted lover! There were few Werthers in those days. Bad poets, and bores of all sorts, were

Early and late creative eras.

not likely to flourish in a society where ostracism, the custody of the Eleven, and the draught of hemlock were looked upon as rather mild and exemplary modes of criticism.

Now, in distinction from unsophisticated and creative song, comes the voice of the poet absorbed in his own emotions and dependent on self-analysis for his knowledge of life. Here is your typical modern minor poet. But here also are some of the truest "bards of passion and of pain" that the world has known. Again, there are those who are free from the Parnassian egoism, but whose manner is so pronounced that every word they utter bears its author's stamp: their tone and style are unmistakable. Finally, many are confined implacably to certain limits. One cares for beauty alone, an artist pure and simple; another is a balladist; a third is gifted with philosophic insight of nature; still another has a genius for the psychological analysis of life. Each of these appears to less advantage outside his natural range. The vision of all these classes is conditioned.

The subjective poet.

Specialists.

An obvious limitation of the speechless arts is that they can be termed subjective only with respect to motive and style. We have the natural landscapist, and the figure-painter, while nearly all good painters, sculptors, architects, musicians, are recognizable, as you know, by their respective styles, but otherwise all arts, save those of language, are relatively impersonal and objective.

Language the freest means of self-expression.

The highest faculties of vision and execution are required to design an absolutely objective poem, and to insure its greatness. There is no middle ground; it is great, or else a dull and perfunctory mechanism. The force of the heroic epics, whose authorship is in the crypt of the past, seems to be not that of a single soul, but of a people; not that of a generation, but of a round of eras. Yet the final determination of poetic utterance is toward self-expression. The minstrel's soul uses for its medium that slave of imaginative feeling, language. It is a voice — a voice; and the emotion of its possessor will not be denied. The poet is the Mariner, whose heart burns within him until his tale is told:

> "I pass, like night, from land to land;
> I have strange power of speech;
> That moment that his face I see,
> I know the man that must hear me:
> To him my tale I teach."

Races themselves have a bent toward one of the two generic types, so that with one nation or people the creative poet is the exception, and with another the rule. *Racial tendency.* The Asiatic inspiration, even in its narrative legacies, is more subjectively vague than that which we call the antique — that of the Hellenes. But the extreme *Asia.* Eastern field requires special study, and is beyond the limits of this course, so that I will only confess my belief that much of our fashionable adaptation of Hindu, Chinese, Japanese literatures represents more honestly the ethics and poetic spirit of its Western students than the Oriental feeling and conceptions; that it is a latter-day illumination of Brahmanic esoterics rather than the absolute Light of Asia,—whether better or worse, not a veritable transfer, but the ideal of Christendom grafted on the Buddhist stock. It is doubtful, in fact, whether the Buddhists themselves fully comprehend their own antiquities; and if our learned virtuosos, from Voltaire and Sir William Jones to Sir Edwin Arnold, fail to do so, they nevertheless have found the material for a good deal of interesting verse. It will be a real exploit when some one does for the Buddhist epos and legendary what John Payne and Captain Burton have done for the Arabic "Thousand Nights and One Night." Then we shall at least know those literatures as they are; nor will it be strange if they prove to be, in some wise, as much superior to our conception of them as Payne's rendering of the "Arabian Nights" is to that of Galland, or as Butcher and Lang's prose translation of Homer is to Lord Derby's verse. Of such a paraphrase as Fitzgerald's "Rubáiyát of Omar Khayyám," one at once declares, in Landor's phrase, that it is more original than the originals: the Western genius in this instance has produced an abiding poem, unique in its interfusion of the Persian and the neo-English dispositions.

But with Hebrew poetry, that of the Bible, we

Attempts to transfer the Buddhist conceptions.

have more to do, since we derive very closely from it. There is no literature at once so grand and so familiar to us. Its inherent, racial genius was emotional and therefore lyrical (though I am not with those who deem all lyrical poetry subjective), and a genius of so fiery and prophetic a cast that its personal outbursts have a loftiness beyond those of any other literature. The Hebrew was, and the orthodox Israelite remains, a magnificent egoist. Himself, his past, and his future, are a passion. But — and this is what redeems his egoism — they are not his deepest passion; he has an intenser emotion concerning his own race, the chosen people, a more fervent devotion to Jehovah, — his own Jehovah, if not the God of a universe. Waiving the question whether the ancient Jew was a monotheist, we know that he trusted in the might of his own God as overwhelmingly superior to that of all rivals. His God, moreover, was a very human one. But the Judaic anthropomorphism was of the most transcendent type that ever hath entered into the heart of man. *The Hebraic genius.*

I do not, then, class the Hebrew poetry, which, though lyrical, gives vent not so much to the self-consciousness of the psalmist or prophet or chieftain as to the pride and rapture of his people, with that which is personal and relative, any more than I would count the winged Pindar in his splendid national odes, or even his patriotic Grecian followers, as strictly subjective, however lyrical *Its national exaltation.*

and impassioned. Such bards are trumpet-tongued with the exaltation of their time and country: they speak not of themselves, but for their people. To the burning imagination of Moses and the prophets, and to the rhythmical eloquence of the Grecian celebrants, I may refer when noting the quality of inspiration. I think the national and religious utterance of the Hebrews even more characteristic than their personal outgivings; they were carried out of and Intense personal feeling. above themselves when moved to song. But there is no more wonderful poetry of the emotional order than the psalms of David and his compeers relating to their own trials and agonies, their loves and hates and adoration. As we agonize and triumph with a supreme lyrical nature, its egoism becomes holy and sublime. The stress of human feeling is intense in such poetry as that of the sixth Psalm, where the lyrist is weary with groaning, and waters the couch with his tears, exclaiming, " But thou, O Lord, how long ? " and that of the thirteenth, when he laments : " How long wilt thou forget me, O Lord ? Forever ? " and in successive personal psalms wherein the singer, whether David or another, avows his trust in the Deity, praying above all to overcome his enemies and to have his greatness increased. These petitions, of course, do not reach the lyrical splendor of the psalms of praise and worship : " The heavens declare the glory of God," " The earth is the Lord's and the fulness thereof ; " and those of Moses — " He that dwelleth in the secret place of the Most High," and its immediate successors. But the Hebrew, in those strains where he communes with God alone, other protectors having failed him, is at the climax of emotional song.

Modern self-expression is not so direct and simple. We doubt the passion of one who wears his heart upon his sleeve. The naïveté of the Davidic lyre is beyond question, and so is the superb unrestraint of the Hebrew prophecy and pæans. We feel the stress of human nature in its articulate moods. This gives to the poetry of the Scriptures an attribute possessed only by the most crea- Subjective, yet universal. tive and impersonal literature of other tongues,— that of universality. Again, it was all designed for music, by the poets of a musical race; and the psalms were arranged by the first composers, — the leaders of the royal choir. It retains forever the fresh tone of an epoch when lyrical composition was the normal form of expression. Then its rhythm is free, unrestrained, in extreme opposition to that of classical and modern verse, relying merely upon antiphony, alliteration, and parallelism. Technical abandon, allied with directness of conception and faithful revelation of human life, makes for universality; makes of the Hebrew Scriptures a Bible, a world's book that can be translated into all tongues with surpassing effect, notably into a language almost as direct and elemental as its own, that of our Anglo-Saxon in its Jacobean strength and clarity.

Advancing further, you perceive that where a work survives as an exception to the inherent temper of a people, it is likely to exhibit greatness. The sublimest poem of antiquity is impersonal, yet written in the Hebrew tongue. The book of Job, the life-drama of the Man of Uz, towers with no peak near it; its authorship lost, but its fable associated in mind with the post-Noachian age, the time when God discoursed with men and the stars hung low in the empyrean. It is both epic and dramatic, yet embodies the whole wisdom of the patriarchal race. Who composed it? Who carved the Sphinx, or set the angles of the Pyramids? The shadow of his name was taken, lest he should fall by pride, like Eblis. The narrative prelude to Job has the direct epic simplicity, — a Cyclopean porch to the temple, but within are Heaven, the Angels, the plumed Lord of Evil, before the throne of a judicial God. The personages of the dialogue beyond are firmly distinguished: Eliphaz, Bildad, Zophar, Elihu,— to whom the inspiration of the Almighty gave understanding, — and the smitten protagonist himself, majestic in ashes and desolation. Each outvies the other in grandeur of language, imagination, worship. Can there be a height above these lofty utterances? Yes; only in this poem has God answered out of the whirlwind, his voice made audible, as if an added range of hearing for a space enabled us to comprehend the reverberations of a superhuman tone. I

The Book of Job.

speak not now of the motive, the inspiration, of the symphonic masterpiece; it is still a mortal creation, though maintaining an impersonality so absolute as to confirm our sense of mystery and awe.

It has been said of the Hebrew language that its every word is a poem; and there are books of the Old Testament, neither lyrical nor prophetic, so exquisite in kind that I call them models of impersonal art. Considered thus, the purely narrative idyls of Esther and Ruth have so much significance that I shall have occasion to recur to them with reference to poetic beauty and construction.

Turning from Semitic literature to the Aryan in its Hellenic development, we at once *Greece.* enter a naturally artistic atmosphere. Until after his Attic prime, the Greek, with no trick of introspection, concerned himself very little about his individual pathology, being far too much absorbed with an inborn sense of beauty, and with his office of imaginative creation. His great lyri- *The lyrists.* cal poets — Alcæus, Simonides, Pindar — rehearsed, as I have said, the spirit of a people rather than of themselves. As with the Hebrews, but conversely, the few exceptions to this usage were very notable, else they could not have arisen at all. One extremity of passion for which, in their sunlit life, they found expression compulsive, was that of love; and among those who sang its delights, or lamented its incompleteness, we have the world's accepted type in

Love's priestess of Mitylene, the "violet-crowned, pure, sweetly smiling Sappho." The pity of it is

Sappho. that we have only the glory of her name, celebrated by her contemporaries and successors, and justified to us by two lyrics in the stanzaic measure of her invention, and by a few fragments of verse more lasting than the tablets of the Parthenon. But the "Hymn to Aphrodite" and the Φαίνεταί μοι κῆνος are enough to assure us that no other singer has so united the intensity of passion with charm of melody and form. A panting, living woman, a radiant artist, are immanent in every verse. After twenty-five centuries, Sappho leads the choir of poets that have sung their love; and from her time to that of Elizabeth Browning no woman has so distinguished her sex. The Christian sibyl moved in a more ethereal zone of feeling, but could not equal her Ægean prototype in unerring art, although, by the law of true expression, most artistic where she is most intense.

The note which we call modern is frequent in the

Classical expressions of feeling. dramas of Euripides, and in those of his satirist, Aristophanes; it drifts, in minor waves of feeling, with the lovely Grecian epitaphs and tributes to the dead, — that feeling, the breath of personal art, which Mahaffy illustrates from the bas-reliefs and mortuary emblems which beautify the tombs west of Athens. The Greek anthology is, rich with sentiment of this cast, so pathetic — and so human. As an instance of what I mean, let me repeat Cory's imitation of the elegiacs of Callimachus on his friend Heracleitus :—

> "They told me, Heracleitus, they told me you were dead, —
> They brought me bitter news to hear and bitter tears to shed.
> I wept, as I remembered how often you and I
> Had tired the sun with talking and sent him down the sky.
>
> "And now that thou art lying, my dear old Carian guest,
> A handful of gray ashes, long, long ago at rest,
> Still are thy pleasant voices, thy nightingales, awake,
> For Death — he taketh all away, but them he cannot take."

This, to be sure, is a paraphrase, yet it conveys the feeling better than the more compact version by the poet-scholar Andrew Lang. Nothing can exceed, in its expression of the spirit, Mr. Lang's handling of Meleager's verses to the memory of his loved and lost Heliodora : —

> "Tears for my lady dead,
> Heliodore!
> Salt tears, and strange to shed,
> Over and o'er."

But I quote no more of this melody, since you can find it, in a certain romance of "Cleopatra," shining by contrast with much of that story like the "jewel in an Ethiope's ear." Others of Mr. Lang's elusive, exquisite renderings, done as it seems by the first touch, are incomparable with any lyrical exploits of their kind since "Music's wing" was folded in the dust of Shelley.

Follow the twilight path of elegiac verse to the Alexandrian epoch, and you find the clear Athenian strain succeeded by a compound of artifice and nature, so full of sentiment withal as to seem the forerunner of Christian art, — in some respects the prototype of our own idyllic poetry. The studiously impassioned lament of Moschus for Bion is nearer than the poetry of his dead master, and of that master's master, Theocritus (always excepting the latter's "Thalysia"), to our own modes of feeling and treatment. It set the key for our great English elegies, from Spenser's "Astrophel" and Milton's "Lycidas" to Shelley's "Adonais" and Arnold's lament for Clough. The subjectivity of the Greek idyllists is thus demonstrated. They were influenced largely by the Oriental feeling, alike by its sensuousness and its solemnity, and at times they borrowed from its poets,— as in the transfer by Moschus of a passage from Job into his Dorian hexameters, of which I will read my own version:—

The Greek idyllists.

"Even the mallows — alas! alas! when once in the garden
They, or the pale-green parsley and crisp-growing anise, have perished,
Afterward they will live and flourish again at their season;
We, the great and brave, and the wise, when death has benumbed us,
Deaf in the hollow ground a silent, infinite slumber
Sleep: forever we lie in the trance that knoweth no waking."

We pass with something like indifference to the Latin poets, because their talent, in spite of many noble legacies bequeathed us, so lacked the freedom, the originality, the inimitable poetic subtilties which animated everything that was Grecian. Hellas was creative of beauty and inspi-

Latin sentiment.

ration; Italia, too, was a creative soil, but of government, empire, law. Her poetry, as it was less an impulse and more a purpose, belongs largely to the mixed class. In its most objective portions there is an air of authorship and self-expression. I will not speak now of Lucretius, who sends out the one dauntless ray of contemplative splendor between the Hebraic sages and the seers of our new dispensation. But Vergil is a typical example of the poet whose *style* is so unmistakable that every verse overflows with personal quality, — a style that endures, establishes a pupilage. Vergil borrowed fire from Greece to light the altars of beauty in a ruder land. The Iliad and Odyssey kindled the invention and supplied the construction of his Æneids; the Georgics, his sturdiest cantos, took their motive from Hesiod; the Eclogues are a paraphrase upon Theocritus. But the Mantuan's style is preëminently his own, — the limpid, liquid, sweet, steadfast Vergilian intonation on which monarchs and statesmen hung enchanted, and which was confessedly the parent-voice of many an after bard. Tennyson, in point of a style whose quality is the more distinct for its diffusiveness, — whose potency, to borrow the homœopathic term, is the greater for its perfect trituration, — has been the English Vergil of our day. Browning's trade-mark is, plainly, the antithesis of what I here mean by style. Our own Longfellow furnishes the New World counterpart of Vergil. In the ascetic

[marginal notes: Vergil. His modern countertypes.]

and prosaic America of his early days he excited a feeling for the beautiful, borrowing over sea and from all lands the romance-forms that charmed his countrymen and guided them to taste and invention. His originality lay in the specific tone that made whatever Longfellow's sweet verse rehearsed a new song, and in this wise his own. Mentioning these leaders of to-day only to strengthen my reference to Vergil, — and as illustrating Schlegel's point that "what we borrow from others, to assume a true poetical shape, must be born again within us," — I may add that there is a good deal of per-
Ovid, Catullus, etc. sonal feeling and expression in the Latin epigrammatists and lyrists. We have Ovid with his Tristia of exile, and Catullus with his Sapphic grace and glow, and a Latin anthology of which the tenderest numbers are eloquent of grief for lover and friend gone down to the nebulous pagan underworld. The deaths that touched them most were those of the young and dear, cut off with their lives unlived, their promise of grace and glory brought to naught. Both the Greeks and the Latins, in their joy of life, strongly felt the pathos of this earthly infruition. That famous touch of Vergil's, in the
A touch of nature. sixth Æneid, was not all artifice: the passage in which Æneas sees a throng of shades awaiting their draught of Lethe and reincarnation in the upper world, — and among them the beauteous youthful spirit that in time will become Marcellus, son of the Emperor's sister Octavia,

and heir to the throne of the Cæsars. Every schoolboy, from the poet's day to the present, knows how this touch of nature made Vergil and his imperial listeners kin:—

"Heu, miserande puer! si qua fata aspera rumpas,
Tu Marcellus eris. Manibus date lilia plenis,
Purpureos spargam flores."[1]

From the consecrating beauty of the Latin verse, in a new world and after nineteen centuries, is derived the legend—*Manibus date lilia plenis*—of an American hymn for Decoration Day. Out of the death of a youth as noble and gracious,[2] in whom centred limitless hopes of future strength and joy, the spirit of poetry well may spring and declare—as from yonder tablet in this very place[3]—that his little life was not fruitless, and that its harvest shall be perennial.

A passing reference may be made at this point to a class of verse elegantly produced in various times of culture and refinement: the hearty overflow of the taste, philosophy, good-fellowship, especially of the temperament, of its immediate maker. Thus old Anacreon started off, that Parisian of Teos. When you come to the Latin Horace, who like Vergil took his models from the Greek, you have, above all, the man himself before you: the

The Horatii.

---

[1] "Ah, dear lamented boy! if thou canst break fate's harsh decrees, thou wilt be our own Marcellus. Bring lilies in handfuls; let me strew the purple flowers!"

[2] Percy Græme Turnbull: born May 28, 1878; died February 12, 1887.

[3] Levering Hall, Johns Hopkins University.

progenitor of an endless succession, in English verse, of our Swifts and Priors and Cannings and Dobsons, of our own inimitable Holmes. There are feeling and fancy, and everything wise and witty and charming, in the individuality of these Horatii; they give us delightful verse, and human character in sunny and wholesome moods. One secret of their attractiveness is their apt measurement of limitations; they have made no claim to rank with the great imaginative poets who supply our loftier models and illustrations.

Return for a moment to that creative art which is found in early narrative poetry and the true drama. The former escapes the pale cast of thought through the conditions of its formation and rehearsal. Primitive ballads have a straightforward felicity; many of them a conjuring melody, as befits verse and music born together. Their gold is virgin, from the rock strata, and none the better for refining and burnishing. No language is richer in them than the English. Our traditional ballads, such as "Clerk Saunders," "Burd Ellen," "Sir Patrick Spens," "Chevy-Chace," "Edward! Edward!" usually are better poetry than those of known authorship. Not until you come to Drayton's "Agincourt" do you find much to rival them. What I say applies to the primitive ballads of all nations. Touch them with our ratiocination, and their charm vanishes. The epos evolved from

<small>Absolutely creative song.</small>

<small>Primitive ballads.</small>

such folk-songs has the same directness. The rhythm of its imagery and narrative, swift and strong and ceaseless as a great river, would be sadly ruffled by the four winds of a minstrel's self-expression,— its current all set back by his emotional tides, Epic masterpieces.

> "The hate of hate, the scorn of scorn,
> The love of love."

The modern temper is not quick to apprehend a work of simple beauty and invention. It presupposes, judging from itself, underlying motives even for the legends and matutinal carols of a young people. Age forgets, and fails to understand, the heart of childhood; we "ancients of the earth" misconceive its youth. We even class together the literatures of races utterly opposed in genius and disposition. Some would put the Homeric epos on the same footing with the philosophi- The Homeric epos. cal drama of Job, the end of which is avowedly "to justify the ways of God to men." Professor Snider, who has exploited well the ethical scheme of "Faust," would similarly deal with the Iliad and the Odyssey. Homer, he thinks, had in mind a grand exposition of Providence, divine rule, the nature of good and evil, and so forth, in relation to which the narrative and poetry of those epics are subordinate and allegorical. But why should we reason too curiously? Both instinct and common sense are against it. Whether the Homeric epos was a growth, or an originally synthetic creation, I believe that the legends of the glorious Ionian verse were recited for the delight of telling and hearing; that the unresting, untiring, billowy hexameters were intoned with the unction of the bard; that they do convey the ancestral reverence, the religion, the ethics, of those adventurous dædal Greeks, but simply as a consequence of their spontaneous truth and vitality. Their poets sang with no more casuistic purpose than did the nightingales in the grove of Colonos. Hence their directness, and their unconscious transmission of the Hellenic system of government and worship. If you wish instruction, everything is essentially natural and true. A perfect transcript of life — the best of teachers — is before us. In the narrative books of the Bible the good and bad appear without disguise. All is set forth with the frankness that made the heart of the Hebrew tent-dweller the heart of the world thereafter. In Homer, the deities are *dramatis personæ*, very human, with sovereign yet terrestrial passions; they dwell like feudal lords, slightly above their dependents, alternating between contempt for them and interest in their affairs. But where is the healthy man or boy who reads these epics without an absorption in their poetry and narrative that is the clew to their highest value? I have little patience with the critics who would disillusionize us. What is the use of poetry? Why not, in this workaday world, yield ourselves to its enjoyment? Homer makes us forget ourselves because he is so self-forgetful. He accepts unquestioningly things as they are. The world has now grown hoary with speculation, but at times, in art as in religious faith, except ye be as children ye cannot enter into the kingdom. We go back to the Iliad and the Odyssey, to the creative romance and poesy of all literatures, as strong men wearied seek again the woods and waters of their youth, for a time renewing the dream which, in sooth, is harder to summon than to dispel. Such a renewal is worth more than any moral, when following the charmed wanderings of the son of Laertes, by isle and mainland, over the sea whose waters still are blue and many-voiced, but whose mystic nymphs and demigods have fled forever; it is worth more than a philosophy,

> "When the oars of Ithaca dip so
>   Silently into the sea
> That they wake not sad Calypso,
>   And the hero wanders free.
> He breasts the ocean furrows
>   At war with the words of fate,
> And the blue tide's low susurrus
>   Comes up to the Ivory Gate."

The dramas of the Attic prime, although equally objective with these epics, are superb poetry, with motives not only creative but distinctly religious and ethical. They recognize and illustrate the eternal law which brings a penance upon somebody for every wrong, the in-

The Greek Dramatists.

scrutable Nemesis to which even the Olympian gods are subject. In this respect the "Prometheus Bound," deathless as the Titan himself, is the first and highest type of them all. The chorus, the major and minor personages, the prophetic demigod, and even the ruthless Zeus, take for granted the power of a righteous Destiny. The wrong-doer, whether guilty by chance or by will, as in the case of the "Œdipus Tyrannus" of Sophocles, even pronounces and justifies his own doom. I will not now consider the grandeur of these wonderful productions. Through the supremer endurance of poetry they have come down to us, while the pictures of Zeuxis and Apelles, and the "Zeus" and "Athene" of Pheidias, are but traditions of "the glory that was Greece." The point I make is that Their absolute quality. these are absolute dramas. They are richly freighted, like Shakespeare's, with oracles and expositions; but their inspired wisdom never diverts us from the high inexorable progress of the action. It is but a relief and an adjuvant. You may learn the bent of the dramatist's genius from his work, but little of his own emotions and experiences. Nor is the wisdom so much his wisdom, as it is something residual from the history and evolution of his people. The high gods of Æschylus and Sophocles. Æschylus and Sophocles for the most part sit above the thunder: but the human element pervades these dramas; the legendary demigods, heroes, *gentes*, that serve as the personages, — Hermes, Herakles, the houses of Theseus, Atreus, Jason, — all are types of humankind, repeating the Hebraic argument of transmitted tendency, virtue, and crime, and the results of crime especially, from generation to generation. The public delight in the Athenian stage was due to its strenuous dramatic action at an epoch when the nation was in extreme activity. Its religious cast was the quintessence of morals derived from history, from the ethics of the gnomic and didactic bards, from the psychological conditions following great wars and crises such as those which terminated at Salamis and Platæa. Æschylus and Sophocles were inspired by their times. They soared in contemplation of the life of gods and men : no meaner flight contented them. The apparent subjectivity of Euripides is due to his relative modernness. No literature was ever so swift to run its course as the Attic drama, from the Cyclopean architecture of the "Prometheus" to the composite order of "Alcestis" and "Ion." Euripides, freed somewhat from the tyranny of the colossal myths, <small>Euripides.</small> was almost Shakespearian in his reduction of them to every-day life with its vicissitudes and social results. His characters are often unheroic, modern, very real and emotional men and women. Aristophanes, still more various, and at times <small>Aristophanes.</small> equal to the greatest of the dramatists, as a satirist necessarily enables us to judge of his own taste and temper; but in his travesties of the immediate life of Athens he is no more self-intrusive than Molière, twenty centuries later, in his portraits of Tartuffe and Harpagon and " Les Précieuses." Men create poetry, yet sometimes poetry creates a man for us, — witness our ideal of the world's Homer. The hearts of the Grecian dramatists were so much in their business (to use the French expression) that they have told us nothing of themselves; but this implies no insignificance. So reverse to commonplace, so individual were they each and all, that in point of fact we know from various sources more of their respective characters, ambitions, stations, than we know of that chief of dramatists who was buried at Stratford less than three centuries ago.

But I well may hesitate to discourse upon the Greek and Latin poets to the pupils of an admired expounder of the classical literatures;[1] and I use the word "literatures" advisedly, since, with all his philological learning, it is perhaps his greatest distinction to have led our return to sympathetic comprehension of the style and spirit of the antique masters, — to have applied, I may say, his genius not only to the materials in which they worked, but to the grace and power and plenitude of the structures wrought from those materials. With less hesitation, then, I change, in quest of strictly dramatic triumphs, from the time of Pericles to the period of Calderon, of Molière, of Shakespeare and his Elizabethan satellites. Lowell says that

Tribute to an American scholar.

[1] Professor Basil L. Gildersleeve.

Addison and Steele together made a man of genius. Terence and Plautus between them perhaps display the constituents of a master-playwright, but not, I think, of a strongly imaginative poet.

I have alluded to the process by which the epic and dramatic chieftains appear to reach their creative independence. As a preliminary, or at certain intervals of life, they seem to rid themselves of self-consciousness by its expression in lyrics, sonnets, and canzonets. Of this the minor works of Dante, Tasso, Boccaccio, Michelangelo, Cervantes, Calderon, Camoëns, Shakespeare are eminent examples. But nothing so indicates the unparalleled success of the last-named poet in this regard, as the fact that, unambiguous as are his style and method, and also his moral, civic, and social creeds, we gather so little of the man's inner and outer life from his plays alone: except as we seem to find all lives, all mankind, within himself, — all experiences,

<blockquote>
"All thoughts, all passions, all delights,
Whatever stirs this mortal frame;"
</blockquote>

and Coleridge, when he called him the myriad-minded, should have added, "because the myriad-lived."

The grand drama, then, like the epic, gives us that "feigned history" which is truer than history as written, because it does not attempt to set things right. Its strength must be in ratio to its imper-

<small>The cry of adolescence. Cp. "Poets of America": p. 146.</small>

sonality. It follows the method of life itself, which to the unthinking so often seems blind chance, so often unjust; and of which philosophers, reviewing the past, are scarcely able to form an ethical theory, and quite helpless to predicate a future. Scientifically, they doubt not,— they must not doubt,— that

The drama an imaginative transcript of life.

> "through the ages one increasing purpose runs,
> And the thoughts of men are widened with the process of the suns."

Right prevails in the end; crime brings punishment, though often to the innocent. We have seen that, if poets, they deal with phenomena, with the shows of things, and, as they see and faithfully portray these, the chances of life seem much at haphazard. Hamlet, for all his intellect and resolve, is the sport of circumstance. Rain still falls upon the just and the unjust. The natural law appears the wind of destiny. Man, in his conflicts with the elements, with tyranny, with superstition, with society, most of all with his own passions, is still frequently overthrown. It *seems* as if the good were not necessarily rewarded except by their own virtue, or, if self-respecting, except by their own pride, holding to the last; the evil are not cast down, unless by their own self-contempt, and the very evil flourish without conscience or remorse. The pull of the universe is upon us, physically as well as morally. When all goes well, and a fair ending is promised, then

Its ethics the law of Nature.

> "Comes the blind Fury with the abhorrèd shears,
> And slits the thin-spun life."

Thus Nature, in her drama, has no temporary pity, no regret. She sets before us the plots of life, and its characters, just as they are. The plots may or may not be laid bare; the characters often reveal themselves in speech and action. As the stream rises no higher than its fount, the ideal dramatist is not more learned than his teacher. He may know no more than you of his personages' secrets. Thackeray confessed, you remember, that Miss Sharp was too deep for him.

Tragedy, according to Aristotle and in Dryden's English, is "an imitation of one entire, great, and probable action, not told but represented, which, by moving in us fear and pity, is conducive to the purging of those two passions in our minds." *Why tragedy elevates the soul.* And so its reading of the book of life, even with our poor vision, is more disciplinary, more instructive in ethics and the conduct of life, than any theoretic preachment. The latter will be colored, more or less, by the temper of the preacher. Besides, through the exaltation to which we are lifted by the poet's large utterance, our vision is quickened: we see, however unconsciously, that earthly tragedies are of passing import, — phenomenal, formative experiences in the measureless progress of the human soul; that life itself is a drama in which we are both spectators and participators; that, when the curtain falls, we may wake as from a dream, and enter upon a life beyond terrestrial tragedies and which fears not even a disembodied phantom, "being a thing immortal as itself."

The Greeks conceived their gods to be almost as powerless as a human protagonist to divert the tides of circumstance, and postulated a Destiny above them all. The dramatists of Christendom, while also impelled to treat life as it is, its best and its worst, recognize no conflict between Deity and Destiny. Pagan and Christian alike present man, the image of his Maker, as exercising his highest·function when he rises superior to fate. Thus Job rises, and thus rise Prometheus, Œdipus, Brutus, Hamlet, Wallenstein, Faust, Van Artevelde, and Gregory VII.; and likewise their fine heroic countertypes, Electra, Alcestis, Antigone, Cordelia, Desdemona, Thekla, Jeanne D'Arc, Doña Sol, and all the feminine martyrs of the grand drama.

<small>Man's victory over fate.</small>

In arguing that the strength of a play is in ratio to its objectivity, I assume, of course, that other things are equal. After all, the statements are the same, for only the poet endowed with insight and passion *can* give a truthful, forcible transcript of life. Otherwise many would outrank Shakespeare, being equally impersonal, more artistic in plot-structure, truer perhaps to history and to the possibilities of events. They often compose successful plays, striking as to incident and use of stage accessories: but more is required — the imagination that creates brave personalities, the cognate high poetic gifts — to make a composition entirely great. Add to such endowments the faculty of self-efface-

<small>The dramatic genius.</small>

ment, and Shakespeare stands at the head thus far. His period fitted him, — one of action and adventurous zest rather than of introspection. At that time, moreover, literary fame and subsistence were won by play-writing. His mind caught fire by its own friction, as he wrote play after play directly for the stage, knowing himself to be in constant touch with the people for whom and from whom he drew his abundant types.

I have often thought upon the relative stations of the various classes of poetry, and am disposed to deem eminence in the grand drama the supreme eminence; and this because, at its highest, the drama includes all other forms and classes, whether considered technically or essentially. Its plot requires as much inventive and constructive faculty as any epic or other narrative. Action is its glory, and characterization must be as various and vivid as life itself. The dialogue is written in the most noble, yet flexible measure of a language; if English, in the blank verse that combines the freedom of prose with the stateliness of accentual rhythm. The gravest speech, the lightest and sweetest, find their best vehicle in our unrhymed pentameter; again, a poetic drama contains songs and other interludes which exercise the lyrical gift so captivating in the works, for example, of our English playwrights: the Elizabethans having been lions in their heroics, eagles in their wisdom, and

Grand drama the noblest and most inclusive of poetic structures.

skylarks in their rare madrigals and part-songs. Tragedy and comedy alike are unlimited with respect to contrasts of incident and utterance, light and shadow of experience; they embrace whatsoever is poetic in mirth, woe, learning, law, religion — above all, in passion and action. So that the drama is like a stately architectural structure; a cathedral that includes every part essential to minor buildings, and calls upon the entire artistic brotherhood for its shape and beauty: upon the carver and the sculptor for its reliefs and imagery; upon the painter and the decorative artist for its wall-color and stained glass; upon the moulder to fashion its altar-rail, and the founder to cast the bells that give out its knell or pæan to the land about. The drama is thus more inclusive than the epic. There is little in Homer that is not true to nature, but there is no phase of nature that is not in Shakespeare.

Analyze the components of a Shakespearian play, and you will see that I make no overstatement.

"The Tempest," a romantic play, is as notable as any for poetic quality and varied conception. It takes elemental nature for its scenes and background, the unbarred sky, the sea in storm and calm, the enchanted flowery isle, so

<small>"The Tempest" as an illustration.</small>

> "full of noises,
> Sounds and sweet airs that give delight and hurt not."

The personages comprise many types, — king, noble, sage, low-born sailor, boisterous vagabond, youth and maiden in the heyday of their innocent love. To them are superadded beings of the earth and air, Caliban and Ariel, creations of the purest imagination. All these reveal their natures by speech and action, with a realism impossible to the tamer method of a narrative poem. Consider the poetic thought and diction : what can excel Prospero's vision of the world's dissolution that shall leave "not a rack behind," or his stately abjuration of the magic art? Listen, here and there, to the songs of his tricksy spirit, his brave chick, Ariel: "Come unto these yellow sands," "Full fathom five thy father lies," "Where the bee sucks, there suck I." Then we have a play within a play, lightening and decorating it, the masque of Iris, Ceres, and Juno. I recapitulate these details to give a perfectly familiar illustration of the scope of the drama. True, this was Shakespeare, but the ideal should be studied in a masterpiece ; and such a play as "The Tempest" shows the possibilities of invention and imagination in the most synthetic poetic form over which genius has extended its domain.

For one, I think that Sophocles and Shakespeare have taught us, by example, that greatness in the noblest of poetic structures must *Impersonality of the old masters.* be impersonal. The magician must not directly appear; though, from reflecting upon a Prospero, a Benedick, or a Hamlet, we may guess at certain of his maker's traits ; and in sooth he must know his own heart to read the heart of the world, even while he stands so far aloof that it may be said of him, as of one translated,

> " Far off is he—
> No more subjected to the change and chance
> Of the unsteady planets."

Modern and subjective drama. Yet there is a subjective drama which, as we have learned in our day, is not without greatness derived from the unique genius of its constructor. The poet of England and Italy, whose ashes Venice has so recently surrendered to their shrine in Westminster, doubtless possessed a sturdier dramatic spirit than any Briton since the days of John Webster and John Ford. *Browning.* Browning was a masterful poet in his temper and insight, his flashes of power and passion, his metaphors, and distinguished for his recognition of national and historic types, his acceptance of life, his profound conviction that the system of things is all right, that we can trust it to the end. But his incessant recurrence to this conviction was a personal factor significant of many others. There are numerous and distinct characters in his repertory, but it requires study to apprehend them, for they have but one habit of speech, whatsoever their age or country. *Cp. "Victorian Poets": pp. 294-297, 431-433.* They all indulge, moreover, in that trick of self-analysis which Shakespeare confines to the soliloquies of special personages at critical moments. Even Browning's little maids study their own cases in the spirit of Sordello or Paracelsus. Finally, his whole work is characterized by a strangely individual style and atmosphere. True, it is difficult to mistake an excerpt from Shakespeare at his prime. But why is this? Because Shakespeare's style has unapproachable beauty, strength, flexibility, within the natural method of English verse; his inimitableness is due not to eccentricity, but to a grandeur of quality. His tone, characterization, and dialogue are as varied as nature. Browning's method hardly suggests either our native order of thought or nature's universality. It seems the result of a decision to compose in a peculiar way, but more likely is the honest reflex of his analytic mental processes. That at times it is great, and above that of his contemporaries, must be acknowledged, for his intellect was of a high order.

Swinburne calls his plays "monodramas, or soliloquies of the spirit." The subjectivity which blends their various personages in a common atmosphere does not detract from the effect of his powerful dramatic lyrics and monologues, each the study of a single character. The most striking of these pieces, — their abundance is prodigal, and not one is without excuse for being, — from "My Last Duchess," "Bishop Blougram," "Childe Roland," "Saul," to "A Forgiveness," including nearly all the "Dramatic Lyrics," and "Men and Women," place him among the century's foremost masters. In such studies, and in certain of his dramas, he has created a new type of English poetry that is second only to the Elizabethan. His eminence is taken for granted when we begin to measure him, if only in contrast, by Shakespeare himself: a tribute rendered to scarcely any other

Dramatic lyrics and monologues.

poet save John Keats, and, in that instance, not on the score of mature dramatic quality, but for a diction so prophetic of what in time might be that the world thinks of his youthful shade among the blest as the one permitted to sit at Shakespeare's feet.

I spoke of our sovereign dramatist as being in spirit with his own people, and writing directly for their stage. Browning's earlier plays were written for enactment, and one or two were produced with some success. These, however, to my mind, are not his best work, and his most effective dramas are not, as we say, adapted to stage performance. Yet I rebuke myself, when repeating this cant of the *coulisses*, as I reflect upon the quality that does find vogue with managers and audiences at the present time. Who can predict what will be thought best "adapted to stage performance" when Jove lets down "in his golden chain the Age of better metal" for which Ben Jonson prayed, — the age, at least, of different metal? Even now we follow a grand drama, though it be one of the outlived classical and recitative cast, with absorbed delight, when it is revived by a Salvini. But I believe that Browning himself would have written more and greater dramas, and of an impersonal order, if there had been a theatrical demand for his work after the performances of "Strafford" and "A Blot in the 'Scutcheon." Mischance, and the spirit of the time, may have lost to us a modern Shakespeare. As it is, we have gained a new avatar of dramatic poetry in the works of our Victorian Browning.

The modern stage.

IV.

MELANCHOLIA.

WE have considered ancient poetry, the Hebraic and the classic, from which we so largely derive, finding even in that of the Augustan prime a marked departure from the originative temper of the earlier literatures. Centuries afterward, in Persia, the "Shah Nameh," or Book of Kings, furnished a striking instance of heroic composition: the work of a royal genius, — Firdusi, whose name, signifying Paradise, was given him by the great Mahmoud because he had made that Caliph's court as resplendent as Eden through his epic of "Rustem and Sohrab," his song of "the rise, combats, death"[1] of the Parsee religion and nationality. To produce an epic deliberately that would simulate the primitive mould and manner, in spite of a subjective, almost modern, spirit, seems to have been the privilege of an Oriental, and, from our point of view, half-barbaric, race.

Subjective undertone of later epic masterpieces.

Firdusi.

The strength of the Homeric poems and of the sagas of the North betrays the gladness out of which they sprang, the joy that a man-child is born into

[1] Gosse's Introduction to Miss Zimmern's *Stories Retold from Firdusi.*

the world. They were men-children indeed. Compared with our own recitals, — with even Tasso's "Jerusalem," Ariosto's "Orlando," or the "Lusiad" of Camoëns, — their voice is that of the ocean heard before the sighing of reeds along a river's brim. Nevertheless, we must note that of the few great world-poems the subjective element claims its almost equal share.

<small>Tasso, Ariosto, Camoëns.</small>

As we leave the classic garden, there stands one mighty figure with the archangelic flaming sword. After Dante it may be said that "the world is all before" us "where to choose." Behind him, strive as we may with renaissance and imitation, we need not and cannot return. Heine says that "every epoch is a sphinx which plunges into the abyss as soon as its problem is solved." After a thousand years of the fermentation caused by the pouring in of Christianity upon the lees of paganism, a cycle ended; the shade of Dante arose, and brooded above the deep. From his time there was light again. A climacteric epoch had expired in giving him birth. His own age became Dante, as if by one of the metamorphoses in the "Inferno." And the "Divine Comedy" is equally one with its creator. The age, the poem, the poet, alike are Dante; his epic is a trinity in spirit as in form. Its passion is the incremental heat that serves to weld antique and modern conceptions, the old dispensation and the new.

<small>The "Divina Commedia."</small>

It is said that great poets are always before or behind their ages; Dante was no excep- *Dante.* tion, yet he preëminently lived within his time. Above all else, his epic declares the intense personality that must have voice; not merely expression of the emotion that inspired his minor numbers — themselves enough for fame — addressed to Beatrice, but also of his insight concerning the master forces of human life and faith and the historic turmoil of his era. It was composed when he had matured through knowledge and experience to that ethical comprehension which is the sustaining energy of Job, of the Greek dramatists, of Shakespeare, Milton, and Goethe. Then he cast his spirit, as one takes a mould of the body, in the matrix of the "Divina Commedia." In this self-perpetuation he interpreted his own time as no modern genius can hope to do, — and this is the achievement of personality at its highest. That he might succeed, he was disciplined by controversy, war, grief, exile, until the scales fell from his eyes, and he saw, within the glory of his Church's exaltation, the vice, tyranny, superstition, of that Church at that time, of his people, of his native state. His heart was strengthened for judgment, his manhood for hate, and his vision was set heavenward for an ideal. His epic, then, while dramatically creative, is *The man, the age, and the* at the apex of subjective poetry, doubly *poem.* so from its expression of both the man and the time; hence our chief example of the mixed type, —

that which is compounded of egoism and inventive imagination. Its throes are those of a transition from absolute art to the sympathetic method of the new day.

Dante could effect this only by a symbolism combining the supreme emblems of pagan and Christian schools.

In his allegory of Hell, Purgatory, and, above all, of Paradise, he is the most profound and aspiring of ethical teachers. The feebler handling of symbolism, for art's sake and beauty's, and with an affectation of the virtues, is seen in the "Faërie Queene" of our courtly Spenser, the poet's poet, yet one who never reached the mountain-top of absolute ethics. The tinker Bunyan's similitudes — and he was essentially a poet, writing in English beyond a mere scholar's mastery — are more intrinsically dramatic. But they illustrate a rigid creed, and are below the imagery that sets forth equally human crime and nobleness, the vision that illumines life, churchcraft, statecraft, nationality, art, and religion. Within the eternal blazon of that saturnine bard whose

<small>"On a Bust of Dante." By T. W. Parsons.</small>

> "rugged face
> Betrays no spirit of repose,
>   The sullen warrior sole we trace,
>   The marble man of many woes.
> Such was his mien when first arose
>   The thought of that strange tale divine,
> When hell he peopled with his foes,
>   Dread scourge of many a guilty line.

"War to the last he waged with all
The tyrant canker-worms of Earth;
Baron and duke, in hold and hall,
Cursed the dark hour that gave him birth;
He used Rome's harlot for his mirth;
Plucked bare hypocrisy and crime;
But valiant souls of knightly worth
Transmitted to the rolls of Time."

The antique charm, meanwhile, had fled to England, ever attaching itself to the youth of poesy in each new land. The English springtime!—to be young in it is very heaven, since it is the fairest of all such seasons in all climes. It gladdens the meadows and purling streams of Dan Chaucer's Tales and Romaunts, and in their minstrelsy he forgot himself, like a child that roams afield in May. With Spenser, and the Tudor sonneteers, the self-expressive poetry of England fairly begins. They, and their common antique and Italian models, were the teachers of Milton in his youth. The scholar gave us what is still in the front rank of our English masterpieces and, with one exception, the latest of those rhythmical creations which belong to the world at large. *From Chaucer to Milton.*

Milton in his epic appears less determinedly as the rhapsodist in person than Dante in the "Divine Comedy." He sees his vision by invocation of the Muse, while the Florentine is "personally conducted," one may say, on his tour through the three phantasmal abodes. Doubtless "Paradise Lost" is the more objective work; *"Paradise Lost."*

but with the unparalleled Miltonic utterance, its author's polemic creeds of liberty and religion are conveyed throughout. He also stands foremost among the bards of qualified vision, by virtue of " Samson Agonistes," a classical drama in which he himself indubitably towers as the blind and fettered protagonist.

Milton's early verse is the flower of his passion for beauty and learning, and exquisite beyond that of any young English poet then or now,— his pupil Keats excepted. Had he died after " Il Penseroso," " L' Allegro," and " Lycidas," he would have been mourned like Keats; for their perfection is to-day the model (though usually at second hand) of artists in English verse. In " Lycidas" he freed our rhythm from its first enslavement; its second lasted from Pope's time until the Georgian revival. One mark of the subjectivity of his early poems often has been noted,— they are none too realistic in their transcripts of nature. Milton, as in his greater work, looked inward, and drew his landscape from the Arcadian vistas thus beheld. Besides, he was such a master of the Greek, Latin, and Italian literatures as to be native to their idioms and spirit. His more resolute self-assertion came in argument and song after experience of imposing national events and sore private calamities, when the man was ripe in thought, faith, suffering, and all that makes for character and exaltation. The universe, as he con-

The minor poems of Milton.

His self-expression in the great Puritan epic.

ceived it, was his theme. His hero, the majestic Satan of his own creation, outvies the Æschylean demigod. The Puritan bard, like Dante, idealized an era and a religion. In the matter and style of the sublimest epic of Christendom its maker's individuality everywhere is felt. The blind seer seems dictating it throughout. We see his head bowed upon his breast; we hear the prophetic voice rehearsing its organ-tones; and thus we should see and hear, even if we could forget that outburst at the opening of the Third Book, wherein, after the radiant conception of the "Eternal coeternal beam," the sonorous declaration of his purposed higher flight, and the pathetic references to his blindness, his final invocation enables all after-time to recognize the inward light from which his imagination drew its splendor: —

> "So much the rather thou, celestial Light,
> Shine inward, and the mind through all her powers
> Irradiate, there plant eyes, all mist from thence
> Purge and disperse, that I may see and tell
> Of things invisible to mortal sight."

Milton's eventide sonnets, incomparable for virility and eloquence, are also nobly pathetic; *His sonnets.* there are no personal strains more full of heroic endurance. Not again was there a minstrel so resolved on personal expression, yet so creative, so full of conviction that often begat didacticism, yet so sensitive to impressions of beauty, until we come to Shelley — and his flight, alas! was ended, while, as Arnold says, he was still " beating in the void his luminous wings in vain."

But the nineteenth century, complex through its interfusion of peoples and literatures, and with all history behind it, has developed the typical poetry of self-expression, and withal a new interpretation of life and landscape through the impressionism of its artists and poets. All this began with the so-called romantic movement.

<small>Our modern and characteristic poetry of self-expression.</small>

Kingsley, in his "Hypatia," brings the pagan Goths of the North, fair-haired worshippers of Odin, giants in their barbaric strength, to Christian Alexandria, where they loom above the Greek, the Roman, and the Jew. In time they overran and to some extent blended with the outer world. It is strange how little they affected its art and letters. Not until after the solvent force of Christianity had done its work could the Northern heart and imagination suffuse the stream of classicism with the warm yet beclouded quality of their own tide. Passion and understanding, as Menzel has declared, represent the antique; the romantic — the word being Latin, the quality German — is all depth and tenderness. To comprehend the modern movement, — vague, emotional, transcendental, — which really began in Germany, read Heine on " The Romantic School," of which he himself, younger than Arnim and Goethe, was a luxuriant offshoot.

<small>The Romantic Movement.</small>

It came into England with Coleridge, with Leigh Hunt and Keats, and found its extreme in Byron. Later still, it fought a victorious campaign in France, under the young Hugo and his comrades. In fine, with color, warmth, feeling, picturesqueness, the iridescent wave swept over Europe, and to the Western world, — affecting our own poetry and fiction since the true rise of American ideality. Upon its German starting-ground the imperial Goe- Goethe. the was enthroned, but he has been almost the only universalist and world-poet of its begetting. For he not only produced with ease the lyrics that made all younger minstrels his votaries, but was fertile in massive and purposely objective work. The drama was his life-study, and he sought to be, like Shakespeare, dramatist and manager in one. "Faust," the master-work of our century, is an "Faust." epochal creation. Yet even "Faust" is the reflection of Goethe's experience as the self-elected archetype of Man, and is subjective in its ethical intent and individuality. Still, the master's tranquil, almost Jovian, nature enabled him often to separate his personality from his inventions. This Hugo. more rarely is the case with the only Frenchman comparable to him in scope and dramatic fertility,— superior to him in energy of lyrical splendor. Melodramatic power and imagination are the twin genii of Hugo, and his human passion is intense; but his own strenuous, untamed temperament compels us everywhere, even in his romantic and historic plays.

He was the true creator of modern French literature, for which he furnished a new vocabulary, and he brought France out of her frigid classicism into line with the Northern world. Then came Lamartine, with his sentiment, and Musset and Gautier, — children of Paris and Helen, consecrate from birth to the abandon of emotion and beauty, and equally with Lamartine to the poetry of self-expression.

<small>Other French romanticists.</small>

Long before, in Scotland, a more spontaneous minstrel also had sung out of the fulness of the music born within him, but with a tone that separated him from the choir of purely subjective poets. Burns was altruistic, because his songs were those of his people. In his notes amid the heather, Scotia's lowly, independent children found a voice. It was his own, and it was theirs; he looked out and not in, or, if in, upon himself as the symbol of his kind. Of all our poets, lyric and idyllic, he is most truly nature's darling; his pictures were life, his voice was freedom, his heart was strength and tenderness. Yet Burns,

<small>Burns.</small>

> " who walked in glory and in joy,
> Following his plough, along the mountain-side,"

is not a child of the introspective Muse. Relatively late as was his song, he stands glad and brave among the simple, primitive, and therefore universal minstrels.

No; it is in Byron, with his loftier genius and more self-centred emotions, that we find our main example of voice and vision conditioned by the temperament of their possessor. Objective poetry, being native to the youth of a race before self-torturing sophistry has wrought bewilderment, seemingly should appeal to the youth of an individual. And thus it does, but to the youngest youth, — that of a wonder-loving child, whom the "Iliad" and "Odyssey," or Scott's epical romances, delight, and who can make little of metrical sentimentalism. The world-weary veteran also finds it a refreshment; his arrogance has been lessened, and he has been taught that his griefs and dreams are but the common lot. *Byron the typical subjective poet.*

Yet it is plain that subjective poetry, if sensuous and passionate, strongly affects susceptible natures at a certain stage of immaturity. Now that town life is everywhere, we see the Wertherism of former days replaced by a kind of jejune æstheticism, with its own peculiar affectation of wit and indifference. But to the secluded youth, not yet concerned with action and civic life, subjective poetry still makes a mysterious appeal. Sixty years ago the young poet of the period, consciously or otherwise, became a Childe Harold, among men, "but not of them;" one who had "not loved the world, nor the world" him. He found a mild dissipation in contemplating his fancied miseries, and was a tragic personage in his own eyes, and usually a coxcomb in those of the unfeeling neighborhood. This mock-heroic pose, so often without *The ferment of new wine.*

a compensating gift, was and is due to the novel consciousness of individuality that comes to each and all, — to the over-consciousness of it which many sentimentalists, against a thousand slights and failures, retain by arrested development to the end of their days. At its best, we have poetic sensibility intensified by egotism. Keats understood this Keats and his unflinching self-analysis. clearly, even when experiencing it. In spite of the real tragedy of his career, he manfully outgrew it; his poetry swiftly advanced to the robust and creative type, as he wasted under a fatal illness and even in his heart's despair. And what better diagnosis of a young poet's greensickness than these words from the touching preface to "Endymion"?

"The imagination of a boy is healthy, and the mature imagination of a man is healthy; but there is a space of life between, in which the soul is in a ferment, the character undecided, the way of life uncertain, the ambition thick-sighted: thence proceed mawkishness, and all the thousand bitters which those men I speak of must necessarily taste in going over the following pages."

It was preordained that even this limbo of life should have an immortal voice, and that voice was Byron. Until his time the sturdy English folk had escaped the need of it. This came with a peculiar agitation of the national sentiment. That Byron found his fame, and the instant power to create an audience for his captivating monodrama, restricted him to a single and almost lifelong mood. This was the more prolonged since it was thoroughly in temper with an eager generation. The French Revolution led to a perception of the insufficiency and brutalism of contemporary systems. Rebellion was in the air, and a craving for some escape to political, spiritual, and social freedom. Byron pointed out the paths by land and sea to a proud solitude, to a refuge with nature and art which the blunted public taste had long forgotten, and he sang so eloquently withal that he drew more than a third part of the rising stars of Europe after him. Their leader is the typical bard of self-expression, not only for the superb natural strength, and directness, and passion of a lyrical genius that forces us to bear with its barbaric ignorance of both art and realism, but because he sustained it to the end of his career in a purely romantic atmosphere. This pervades even the kaleidoscopic "Don Juan," the main achievement of his ripest years, strengthened as it is by the vigor of which humor is the surplusage and an easy-going tolerance the disposition. It must always be considered, in so far as his development was arrested, that Byron was a lord, born and bred in the British Philistinism against which his nature protested, and that the protest was continued because the fortress did not yield to assault. And he had no Byron for a predecessor, as an object-lesson in behalf of naturalness and common sense.

Byron and his period.

Shelley, who came and went like a spirit, and whose poetry seemed the aureole of a strayed visitor from some translunary sphere, is even more present to us than Byron, with whom, by the law that brings the wandering moths of nightfall together, his life touched closely during its later years. His self-portrayal is as much more beautiful and poetic than Byron's as it is more truthful, unaffected, — drawn wholly for self-relief. That it had no theatrical motive is clear from internal evidence, and from his biographer's avowal that he had gained scarcely fifty readers when he died. Byron was consciously a soliloquist on the stage, with the whole reading world to applaud him from the auditorium. Again, while nothing can be more poignantly intense than Shelley's self-delineation in certain stanzas of the "Adonais," and throughout "Alastor," selfishness and egotism had no foothold in his nature. He was altruism incarnate. His personal sufferings were emblematic of wronged and baffled humanity. Thus it was that when removed somewhat from the battle-field, and in the golden Italian clime of beauty and song, his art instinct asserted itself; his poetic faculty at once became more absolute, and he produced "The Cenci," "Prometheus Unbound," and shorter lyrical pieces more than sufficient to prove his greatness in essentially creative work. And thus it was, as we have seen, with Keats, who caught by turns the spirits of Greece, of Italy, of the North. Landor did the same, with his "Hellenics," with his "Pericles and

Aspasia," "Pentameron," and "Citation of Shakespeare." But Landor, with the fieriest personal temper conceivable, was, like Alfieri, though of a totally different school, another being when at work, an artist to his fingers' ends. So was Coleridge at times, when he shook himself like Samson: not the subjective brother-in-arms of Wordsworth, but the Coleridge of the imagination and haunting melody and sovereign judgment unparalleled in his time, — Coleridge of "The Ancient Mariner," and "Christabel," and "Kubla Khan," whose loss to the highest field of poetic design is something for which one never can quite forgive theology and metaphysics. Of Wordsworth, the real master of the Victorian self-absorption, I shall speak at another time, with respect to our modern conception of the sympathetic quality of nature. To conclude, the prodigal Georgian school, springing from a soil that had lain fallow for a hundred years, was devoted as a whole to self-utterance, but magnificently so. Of course a reaction set in, and we now complete the more restrained, scholarly, analytic, artistic Victorian period, — a time, I fully believe, of equally imaginative effort, yet of an effort, as we shall see, that usually has taken, so far as concerns dramatic invention, a direction other than rhythmic.

Wordsworth.

The Georgian School.

Meanwhile, Heinrich Heine, of the intermediate generation, and the countryman of Goethe, began, one might say, where Byron left off. His

Heine.

whole song is the legacy of his personal mood, but that was full of restless changes from tears and laughter, from melody and love and tenderness, to scorn and cynicism, and again from agnosticism to faith. In youth, and at intervals until his death, his dominant key was like Byron's, — dissatisfaction, longing, the pursuit of an illusive ideal, the love of love and fame. There was an apparent decline, after disordered years, in Byron's powers both physical and mental. Yet his Greek campaign bade fair to bring him to something better than his best. He had the soldier's temperament. Action of the heroic kind was what he needed, and might have led to the "sudden making" of a still more splendid name. Heine was many beings in one: a Jew by race, a German by birth, a Parisian by adoption, taste, and instinct for the beautiful. His outlook, then, was broader than that of the English poet. His writing was also a revolt, but against the age as that of a Jew, and against contemporary Philistinism as that of an Arcadian. Byron became a cosmopolite; Heine was born one. In the world's theatre he stood behind the scenes of the motley human drama. He wrought its plaint and laughter into a fantastic music of his own, with a genius both sorrowful and sardonic; always like one enduring life as a penance, and suffering from the acute consciousness of some finer existence the clew to which was denied him : —

"Most musical, most melancholy."

> " In every clime and country
>   There lives a Man of Pain,
> Whose nerves, like chords of lightning,
>   Bring fire into his brain:
> To him a whisper is a wound,
>   A look or sneer a blow;
> More pangs he feels in years or months
>   Than dunce-throng'd ages know."

Heine felt, and avowed, that the actual song-motive is a heart-wound, without which " the true poet cannot sing sweetliest." His mocking note, *The mocking note.* which from its nature was not the sanest art, was quickly caught by younger poets, and repeated as if they, too, meant it, and for its air of experience and maturity. With real maturity they usually hastened to escape from it altogether.

I think that the impersonal element in art may be termed masculine, and that there is some- *The major and minor keys of* thing feminine in a controlling impulse to *lyric song.* lay bare one's own heart and experience. This is as it should be: certainly a man's attributes are pride and strength, — strength to wrestle, upon occasion, without speech until the daybreak. The fire of the absolutely virile workman consumes its own smoke. But the artistic temperament is, after all, androgynous. The woman's intuition, sensitiveness, nervous refinement join with the reserved power and creative vigor of the man to form the poet. As those or these predominate, we have the major strain, or the minor appeal for human sympathy and the proffer of it. A man must have a notable gift or a very exalted nature to make people grateful for his confessions. The revelations of the feminine heart are the more beautiful and welcome, because the typical woman is purer, more unselfish, more consecrated, than the typical man. Through her ardent self-revelations our ideals of sanctity are maintained. She may even, like a child, be least self-conscious when most unrestrained in self-expression. Assuredly this was so in the case of the greatest woman-poet the modern world has known. Mrs. Browning's lyrics, every verse sealed with her individuality, glowing with sympathy, and so unconsciously and unselfishly displaying the nobility of her heart and intellect, have made the earth she trod sacred, and her resting-place a shrine. Her impassioned numbers are her most artistic. The "Sonnets from the Portuguese," at the extreme of proud self-avowal, are equal in beauty, feeling, and psychical analysis to any series of sonnets in any tongue,—Shakespeare's not excepted.

Mrs. Browning.

I have alluded to Alfieri. The poets of modern Italy, romantic as they are, still derive closely from the antique, and they have applied themselves considerably to the drama and to the higher lyrical forms of verse. Chafing as they did so long under the Austrian sway, their more elevated odes, as you will see in Mr. Howells's treatise,

National sentiment.

have been charged with "the longing for freedom, the same impulse toward unity, toward nationality, toward Italy." Poetry that has been the voice and force of a nation occupies, as I have said, a middle ground between our two extremes. It has an altruistic quality. The same generous fervor impetuously distinguished the trumpet-tongued lyrics of our Hebraic Whittier, and the unique outgivings of Lowell's various muse, in behalf of liberty and right. Those were " Noble Numbers;" and, in truth, the representative national sentiment — of which ideas of liberty, domesticity, and religion are chief components — pervades the lyrics of our elder American poets from Bryant to Taylor and Stoddard. Whitman's faith in the common people, in democracy strong and simple, has gained him worldwide honor. Subjective as they are, few poets, in any era or country — and historians will come to recognize this clearly — have been more national than our own.

The latest school, with its motto of art for art's sake, has industriously refined music, color, design, and the invention of forms. *Self-conscious technicists.* But its poets and painters show a kind of self-consciousness in the ostentatious preference of their art to themselves, even in their prostration at the feet of "Our Lady of Beauty." Their motive is so intrusive that the result, although alluring, often smacks of artisanship rather than of free and natural art. Their early leaders, such as the young Tennyson and Rossetti in England, and Gautier in France, effected a potent, a charming, a sorely needed restoration of the beautiful. But the Laureate has lived to see another example of his own saying that a good fashion may corrupt the world. The French Parnassiens, the English-writing Neo-Romanticists, are more constructive than spontaneous, and decorative most of all. They have so diffused the technic of finished verse that the making of it is no more noteworthy than a certain excellence in piano playing. They plainly believe, with Schopenhauer, that "everything has been sung. Everything has been cursed. There is nothing left for poetry but to be the glowing forge of words."

<small>Fin de siècle.</small>

This curious, seemingly impersonal poetry, composed with set purpose, finds a counterpart in some of the bewildering recent architecture. How rarely can we say of the architect and his work,

<small>Latter-day verse.</small>

> "He builded better than he knew:
> The conscious stone to beauty grew."

The artist and the builder are too seldom one. The poet just quoted, when on a trip to New Hampshire, found a large building going up in a country town. "Who is the architect?" he said. "Oh, there is n't any architect settled upon as yet," was the reply; "I'm just a-building it, you see, and there's a chap coming from Boston next month to put the architecture into it." So it is with a good deal of our latter-day verse. It does not rise "like an exhalation." It is merely the similitude of the impersonal, and art for the artist's sake rather than for the sake of art. Its one claim to objectivity is, in fact, the lack of any style whatever, except that derived by the rank and file from their study of the chiefs. It is all in the fashion, and all done equally well. Even the leaders, true and individual poets as they have been,—Tennyson, Rossetti, Swinburne, Morris, Sully Prudhomme, Banville, —often have seemed to compose perfunctorily, not from inspired impulse. Read "The Earthly Paradise," that seductive, tranquillizing, prolonged, picturesque rehearsal of the old wonder-tales. Its phantasmagoric golden haze, so often passing into twilight sadness, has veiled the quality of youth in those immortal legends. What is this that Morris fails to capture in his forays upon the "Odyssey," the "Decameron," Chaucer, the "Gesta Romanorum," the "Edda," the "Nibelungen Lied"? Can it never come again? Has it really passed away? Did it wake for the last time in those lusty octosyllabic romances of the Wizard of the North, such as "Marmion" and the "Lay of the Last Minstrel"? Careless, faulty, diffuse as they were, those cantos were as alive as Scotland herself, and fresh with the same natural genius, disdaining to hoard itself, that produced the Waverley novels. If Scott has had no successor, it is doubtless because the age has needed none. We have moved into

Its lack of the creative impersonality.

Morris and Walter Scott: an illustration.

another plane, not necessarily a lower but certainly a different one.

With respect to style, Swinburne is the most subjective of contemporary poets, yet he has made notable successes in dramatic verse, — chief of all, and earliest, the "Atalanta in Calydon," with whose <small>Swinburne.</small> auroral light a new star arose above our horizon. Nothing had been comparable to its imaginative music since the "Prometheus Unbound," and it surpassed even that — for its author had Shelley for a predecessor — in miracles of rhythmic melody. The "Prometheus" surges with its author's appeal from tyranny; "Atalanta" is a pure study in the beautiful, as statuesque as if done in Pentelican marble. Its serene verse, impressive even in the monometric dialogue, its monologues and transcendent choruses, — conceived in the spirit of Grecian art, but introducing cadences unknown before, — all these are of the first order. The human feeling that we miss in "Atalanta" is, on the other hand, a dramatic factor in Swinburne's Trilogy of Mary Stuart. But in his most impersonal <small>The worth and disadvantage of a strongly individual style.</small> work his fiery lyrical gift and individuality will not be suppressed. The noble dramas of Henry Taylor and Hengist Horne are more objective, but cannot vie with Swinburne's in poetic splendor. Now, as you know, this unrivalled voice is instantly recognized in his narrative romances, or in any strophe or stanza of his plenteous odes and songs. The result is that his vogue has suffered. His metrical genius is too specific, too enthralling, to be over-long endured. Thus the distinctive tone, however beautiful, which soonest compels attention, as quickly satiates the public. The subjective poets who restrict their fertility, or who die young, are those whom the world canonizes before their bones are dust.

While, then, a few modern poets, at times as absorbed as Greeks in their work, have Temperament. been strenuously impulsive in temper and in the conduct of life, — among them Alfieri, Foscolo, Hugo, Landor, Horne, and various lights of the art-school from Keats onward, — the artist's temperament usually in the end determines the order of his product: clearly so in such cases as those of Leopardi, James Thomson, Baudelaire, Poe. Sympathetic examination of the poetry will give you the poet. A fine recent instance of an intro- Arnold's conflict with his genius. spective nature overcoming the purpose formed by critical judgment was that of Matthew Arnold. A preface to the second edition of his poems avowed and defended his poetic creed. Reflection upon the antique, and the study of Goethe, had convinced him that only objective art is of value, and that the most of that which is infected with modern sentiment is dilettanteism. Art must be preferred to ourselves. Action is the main thing; more than human dramatic greatness alone saves even Shakespeare's dramas from being weakened by "felicities" of thought and expression. The poet-critic accordingly proffered his two heroic episodes, "Balder Dead" and "Sohrab and Rustum," — both "Homeric echoes," though in their slow iambic majesty violating his own canon that the epic movement should be swift. These are indeed the *tours de force* of intellect and constructive taste. There are fine things in both, but the finest passages are reflective, Arnoldian, or, like the sonorous impersonation of the river Oxus, and the picture of Balder's funeral pyre, elaborately descriptive, and unrelated to the action of the poems. Now, these blank-verse structures are not quite spontaneous; they do not possess what Arnold himself calls the "note of the inevitable." The ancients, doing by instinct what he bade us imitate, had no cause to lay down such a maxim as his, — that the poet "is most fortunate when he most entirely succeeds in effacing himself." They worked in the manner of their time. Schlegel points out that when even the Greeks imitated Greeks their triumph ended. A modern, who does this upon principle, virtually fails to profit by their example. In the end he has to yield. Arnold was beloved by his pupils — by those whom he stimulated as Emerson stimulated American idealists — for the poetry wherein he was in truth most fortunate, that is, in which he most entirely and unreservedly expressed himself; in verse, for the tender, personal, subtly reflective lyrics that seem like tremulous passages from a psychical journal; most of all, perhaps, for those which so convey the spirit of youth, — the youth of his own doubting, searching, freedom-sworn Oxonian group — a group among whom he and Clough, his scholar-gypsy, were leaders in their search for unsophisticated nature and life, in their regret for inaction, their yearning for new light, their belief that love and hope are the most that we can get from this mortal existence. It was Arnold's sensitive and introspective temperament, so often saddening him, that brought his intellect into perfect comprehension of Heine, Joubert, Sénancour, and, doubtless, Amiel. His ca- "Look in thy heart and reer strengthens my belief that the true write." way is the natural one, — that way into which the artist is led by impulse, modified by the disposition of his time. Burns was a force because he was not Greek, nor even English, but Scottish, entirely national, and withal intensely personal. Scott's epics are founded in the true romantic ballads of the North. A few of us read and delight in "Balder Dead;" "Marmion," a less artistic poem, gave pleasure far and wide, and still holds its own. I confess that this again suggests my old question concerning Landor, "Shall not the wise, no less than the witless, have their poets?" and that, whether wise or otherwise, I prefer to read "Balder Dead;" but I have observed that poetry, however admi- A consideration. rable, which appeals solely to a studious class rarely becomes in the end a part of the world's literature. Palgrave, in the preface to "The Golden Treasury," significantly declares that he "has found the vague general verdict of popular Fame more just than those have thought who, with too severe a criticism, would confine judgments on poetry to 'the selected few of many generations.'"

Like Arnold, nearly all his famous peers of the recent composite period have made attractive experiments in the objective and antique fields, though less openly upon conviction. Yet Tennyson and Browning are essentially English and modern, as Emerson, Longfellow, Whittier, are American and New-English, while Lowell's memorable verse is true to the atmosphere, landscape, national spirit, dialect, of his own land, and always true to his ethical convictions. Our minor artists in verse succeed as to simplicity and sensuousness in their renaissance work, but fail with respect to its passion, — for to simulate that requires vigorous dramatic power. The latter is rarely displayed; its substitute is the note of Self. If this be so, let us make the best of it, and furnish striking individualities for some future age to admire, as we admire the creations of our predecessors. At all events, the poet must not dare anything against nature. Let him obey Wordsworth's injunction,

The composite era.

"If thou indeed derive thy light from Heaven,
Then, to the measure of that heaven-born light,
Shine, Poet! in thy place, and be content."

But are there, then, no dramatic works in recent literature? Yes; more than in any former time, if you do not insist upon poetic form and rhythm. While the restriction adopted for these lectures excludes that which is merely inventive composition, you know that prose fiction is now the principal result of our dramatic impulse. The great modern novels are more significant than much of our best poetry. What recent impersonal poem or drama, if you except "Faust," excels in force and characterization "Guy Mannering" and the "Bride of Lammermoor," "Notre Dame de Paris," "Les Trois Mousquetaires," "Père Goriot," "On the Heights," "Dimitri Rudini," "Anna Karénina," "With Fire and Sword," "Vanity Fair," "Henry Esmond," "The Newcomes," "Bleak House," "The Tale of Two Cities," "The Cloister and the Hearth," "Westward Ho!" "Adam Bede," "Romola," "Lorna Doone," "Wuthering Heights," "The Pilot," "The Scarlet Letter," "The Deluge," and other prose masterpieces with which you are as familiar as were the Athenians with the plays of Euripides? More than one of them, it is true, reflects the author's inner life (but so does "Faust"), and is all the more intense for it. The free nature of the novel seems to make subjectivity itself dramatic. Certainly, the individuality of a Bronté, a Thackeray, a Hawthorne, or a Meredith does not lead us to prefer G. P. R. James, or put them on a lower plane than the strictly

The modern creative spirit and its chief mode of activity.

Our prose fiction. Cp. "Poets of America": p. 463.

objective one of De Foe, Jane Austen, Dumas. Our second-rate novels are chiefly mechanical inventions turned off for a market which the modern press has created and is ominously enlarging. However, with such an outlet for the play of the invention which, three centuries ago, spent its strength upon the rhythmical drama, it is no wonder that even our foremost poets look out to rival ranges, with now and then still another peak above them; and these lectures would seem an anachronism were it not that it is a good time to observe the nature of an object when it is temporarily inactive.

Except for this prose fiction superadded to the best poetic achievements of the modern schools, the nineteenth century would not have been, as I believe it to have been, nearly equal in general literary significance (as in science it is superior) to the best that preceded it. It is difficult for critics to project themselves beyond their time; perceiving its shortcomings, they are prone to underestimate what in after time may seem a peculiar literary eminence. To all the splendor of our greatest fiction must be united the romance of the Georgian poetic school and the composite beauty and thought of the Victorian, that this statement may be sound with respect to the literature of our own language. While poetry and fiction both have to do with verities, Mill was not wrong when he said that the novelist gives us a true picture of life, but the poet, the truth of the soul.

The nineteenth century: its literary distinction.

From our survey, after granting that only a few world-poems exhibit the absolute epic and dramatic impersonality, it by no means follows — in spite of common assertion — that the worth of other poetry is determined by an objective standard. The degree of self-expression is of less moment than that of the poet's genius. Subjective work is judged to be inferior, I take it, from its morbid examples. The visits of the creative masters have been as rare as those of national demigods, and ordinary composers fall immeasurably short of their station. We have the perfect form, historical or fanciful impersonations, but few striking conceptions. The result is less sincere, less inevitable, than the spontaneous utterance of true poets who yield to the passion of self-expression.

Objectivity not the chief test of poetic genius.

Yet we have seen that a line can be rather clearly drawn between the pagan and Christian eras, and that there has been a loss. To think of this as a loss without some greater compensation is to believe that modern existence defies the law of evolution and is inferior as a whole to the old; that the soul of Christendom, because more perturbed and introspective, is less elevated than that of antiquity. Contrast the two, and what do we find? First, a willing self-effacement as against the distinction of individuality; secondly, the simple zest of art-creation, as against the luxury of human

The old and new dispensations.

feeling — a sense that nourishes the flame of consolation and proffers sympathy even as it craves it ; —

> " That from its own love Love's delight can tell,
>   And from its own grief guess the shrouded Sorrow;
> From its own joyousness of Joy can sing;
>   That can predict so well
>   From its own dawn the lustre of to-morrow,
> The whole flight from the flutter of the wing."

This sympathy, this divinely human love, is our legacy from the Teacher who read all joys and sorrows by reading his own heart, being of like passions with ourselves,— a process wisely learned by those fortunate poets who need not fear to obey the maxim, " Look in thy heart and write ! "

The Christian motive has intensified the self-expression of the modern singer. That he is subject to dangers from which the pagan was exempt, we cannot deny. His process may result in egotism, conceit, the disturbed vision of eyes too long strained inward, delirious extremes of feeling, decline of the creative gift. Probably the conventual, middle-age Church, with its retreats, penances, ecstasies, was the nursery of our self-absorption and mysticism, the alembic of the vapor which Heine saw infolding and chilling the Homeric gods when the pale Jew, crowned with thorns, entered and laid his cross upon their banquet-table. It is not the wings alone of Dürer's mystic " Melencolia " that declare her to be a Christian figure. She sits among the

Conventual introspection.

Dürer's " Melencolia " as the Muse of Christendom.

well-used emblems of all arts, the ruins of past achievements, the materials for effort yet to come. Toil is her inspiration, exploration her instinct: she broods, she suffers, she wonders, but must still explore and design. The new learning is her guide, but to what unknown lands? The clew is almost found, yet still escapes her. Of what use are beauty, love, worship, even justice, when above her are the magic square and numbers of destiny, and the passing-bell that sounds the end of all? Before, stretches an ocean that hems her in. What beyond, and after? There is a rainbow of promise in the sky, but even beneath that the baneful portent of a flaming star. Could Dürer's "Melencolia" speak, she might indeed utter the sweet and brave, yet pathetic, poetry of our own speculative day.

Our view of the poetic temperament is doubtless a modern conceit. The ancient took life as he found it, and was content. Death he accepted as a law of nature. Desire, the lust for the unattainable, aspiration, regret, — these are our endowment, and our sufferings are due less to our slights and failures than to our own sensitiveness. Effort is required to free our introspective rapture and suffering from the symptoms of a disease. Yet there is no inevitable relation between disease and genius, and it is chiefly in modern song that "great wits are sure to madness near allied." Undoubtedly at feverish crises a flood of wild imaginings overwhelms us. Typical poets have acknowledged this,

Neurotic sensitiveness.

— Coleridge, Byron, Heine, who cite also the cases of Collins, Cowper, Novalis, Hoffmann, and other children of fantasy and sorrow. Coleridge pointed to those whose genius and pursuits are subjective, as often being diseased; while men of equal fame, whose pursuits are objective and universal, the Newtons and Leibnitzes, usually have been long-lived and in robust health. Bear in mind, however, the change latterly exemplified by Wordsworth, Tennyson, Browning, Hugo, and our vigorous American Pleiad of elder minstrels, who have exhibited the sane mind in the sound body. But the question of neurotic disorder did not occur to the age of Sophocles and Pindar. Impersonal effort is as invigorating as nature itself: so much so that Ruskin recognizes the great writer by his guiding us far from himself to the beauty not of his creation; and Couture, a virile figure, avowed that "the decline of art commenced with the appearance of personality." Goethe, in spite of his own theory, admitted that the real fault of the new poets is that "their subjectivity is not important, and that they cannot find matter in the objective." The young poets of our own tongue are not in a very different category. The best critic, then, is the universalist, who sees the excellence of either phase of expression according as it is natural to one's race and period. A laudable subjectivity dwells in naturalness, — the lyrical force of genuine emotions, including those animated by the Zeitgeist of one's

The health of nature.

own day. All other kinds degenerate into sentimentalism.

If we have lost the antique zest, the animal happiness, the naïveté of blessed children who know not the insufficiency of life, or that they shall love and lose and die, we gain a new potency of art in a sublime seriousness, the heroism that confronts destiny, the faculty of sympathetic consolation, and that "most musical, most melancholy" sadness which conveys a rarer beauty than the gladdest joy,— the sadness of great souls, the art-equivalent of the melancholy of the Preacher, of Lincoln, of Christ himself, who wept often but was rarely seen to smile. The Christian world has added the minor notes to the gamut of poesy. It discovers that if indeed "our sweetest songs are those which tell of saddest thought," it is better to suffer than to lose the power of suffering. *[Modern ideality: the loss and gain.]*

Commonplace objective work, then, is of no worth compared with the frank revelation of an inspiring soul. Our human feeling now seeks for the personality of the singer to whom we yield our heart. Even Goethe breaks out with "Personality is everything in art and poetry;" Schlegel declares that "a man can give nothing to his fellow-man but himself;" and Joubert — whom Sainte-Beuve has followed — says, "We must have the man.... It is human warmth and almost human substance which gives to all things that quality which charms us." This fact is a strong- *["To thine own self be true."]*

hold for the true impressionists. The special way in which his theme strikes the artist is his latter-day appeal. And what is style? That must be subjective. Some believe it to be the only thing which is the author's own. The modern mind understands that its compensation for the loss of absolute vision is the increase of types, the extension of range and variousness. These draw us nearer the plan of nature, that makes no two leaves alike. The value of a new piece of art now is the tone peculiar to its maker's genius. Death in art, as in nature, is now the loss of individuality, — a resolution into the elements. We seek the man behind the most impersonal work; more, the world conceives for itself ideals of its poets, artists, and heroes, plainly different from what they were, yet adapted to the suggestions received from their works and deeds.

My summary, then, is that the test of poetry is not by its degree of objectivity. Our inquiry concerns the poet's inspiration, his production of beauty in sound and sense, his imagination, passion, insight, thought, motive. Impersonal work may be never so correct, and yet tame and ineffective. Such are many of the formal dramas and pseudo-classical idyls with which modern literature teems. Go to, say their authors, let us choose subjects and make poems. The true bard is chosen by his theme. Lowell "waits" for "subjects that hunt me." Where the nature of the singer is noble,

The essential rule of judgment.

his inner life superior to that of other men, the more he gives us of it the more deeply we are moved. We suffer with him; he makes us sharers of his own joy. In any case the value of the poem lies in the credentials of the poet.

It is the same with all other speculations upon art: with that, for instance, concerning realism and romanticism, of late so tediously bruited. *Disputed methods in art.* Debate of this sort, even when relating to the Southern and the Wagnerian schools of music, or to impressional and academic modes of painting, is often inessential. It has, perchance, a certain value in stimulating the members of opposing schools. The true question is, How good is each in its kind? How striking is the gift of him who works in either fashion? Genius will inevitably find its own fashion, and as inevitably will pursue it.

V.
BEAUTY.

FOR the moment, and somewhat out of the order of discussion, I will assume that no poem can have birth without that unconscious process of the soul which is recognized in our use of words like "intuition," "insight," "genius," "inspiration." Nor can it be brought to completeness without the exercise of conscious afterthought. True poetry, however, is reinforced by three dynamic elements. No work of art is worth considering unless it is more or less effective through beauty, feeling, and imagination; and in the consideration of art, truth and ethics are a part of beauty's fidelity to supreme ideals. *Poetry as an artistic expression of the beautiful.*

You will find it needful to examine the nature of that which is termed Beauty, before ackowledging that poetry can be no exception, but rather the chief illustration, when it is declared that an indispensable function of the arts is the expression of the beautiful. *What, then, is Beauty?*

With respect to the artists and critics who abjure that declaration, — as when, for instance, a critic said of an American draughtsman that he was too fine an artist to concern himself *The denial of its indispensability.*

about mere beauty, — I am convinced that they simply are in rebellion against hackneyed standards. They have adopted some fresh, and therefore welcome, notion as to what is attractive. This they have given a new name, to distinguish it from an established and too familiar standard. They are unwittingly wooing beauty in a new dress, — the same goddess, with more disguises than Venus Mater. Some day they will recognize her, *et vera incessu patuit dea*, and again be taught that she never permits her suitors to escape. She has the secret of keeping them loyal in spite of themselves. This belief that they are free is a charm by which she lures them to her unknown haunts, rewarding them with the delight of discovery, or ironically permitting them to set up claims for invention. One may even compare beauty to the wise and charming wife who encourages a fickle husband's attentions at a masquerade. She has a thousand graces and coquetries. At last the masks are removed. "What, is it you? And still superior to all others!" He needs must worship her more than ever, and own that none can rival her adorable and "infinite variety."

No; the only consistent revolt is on the part of those who declare that she has no real existence, — that beauty is a chimera. Let me confess at once that I am not in their ranks. I doubt whether any artist, or any thinker who honestly loves art and has an instinct for it,

<small>A logical but purely Berkeleian theory.</small>

believes this theory of æsthetics, though he may advocate it or be driven into its acceptance. An argument can be made on that side, granting certain premises. Even then it is a dispute about terms. The claim may serve for metaphysicians, not for those whose vocations relate to the expression of artistic ideas in what is called tangible form. Go back to Berkeley and his forebears, if you like. Deny the existence of all things, — for that is what you must do if you deny the actuality of beauty, else you are instantly routed. Your only safe claim is that naught but soul exists, and this not the general soul, but your own soul, your Ego. You think, therefore you are; everything else is, for all that I can prove, the caprice of your own dream. Some of our modern transcendentalists, vaunting their Platonic allegiance to ideal beauty, af- *Its supporters.* fected indifference to its material emblems. The modern impressionists, after all the most ardent and ingenuous of technicists, are unwittingly their direct successors. Now, the transcendentalists often were speculators, and not, as they deemed themselves, artists and poets. Having little command over the beautiful, they took refuge in discrediting it. I speak of certain of the followers: their chief was Argus-eyed. In Emerson *Emerson's* the true poet constantly broke loose. He, *own view.* too, looked inward for the ideal beauty, that purest discovery of the soul, but in song he always recognized its visible reality: —

> "For Nature beats in perfect tune,
>   And rounds with rhyme her every rune,
>   Whether she work in land or sea,
>   Or hide underground her alchemy.
>   Thou canst not wave thy staff in air,
>   Or dip thy paddle in the lake,
>   But it carves the bow of beauty there,
>   And the ripples in rhymes the oar forsake."

Impressionism. But, as I say, the recantation of beauty, by transcendentalists, realists, and impressionists alike, is the search for her in some other of her many realms. Whatsoever kingdom the impressionist enters, he still finds her on the throne. For him she may veil herself in twilight and half-tints, — or at rare instants of perception in still more witching drapery worn for him alone. The individual impressions enrich our museum of her portraitures. The impressionist depicts her not as she was known to Pheidias, or Raphael, or Velasquez, but as she appears to his own favored vision. This is the truth that makes impressionism a brave factor in modern art and poetry. What lessens its vantage is the delusion, absurd as Malvolio's, of incompetents, each of whom fancies that he is in special favor and that myopic vision and eccentric technic result in impressions that are worth recording.

Whenever there is a notable break from that mediocrity falsely termed "correct," which lurks in academic arras, it is not a rebellion but a just revolution. This is why it has been said that "the strength of Shakespeare lay in the fact that he had no taste; he was not a man of letters." But men of letters now accept Shakespeare as their highest master. Thus every new movement or method in art has the added form of strangeness at first, — of a true romanticism. In time this, too, becomes classicism and academic. The mediocrities, the dullards of art, are ever the camp-followers of its shining soldiery. In every campaign, under every mode that a genius brings into vogue, they ultimately pitch their ragged tents; and even if they do not sink the cause into disrepute, they make in time a new departure necessary. In the greatest work, however, there will be found always a fresh originality that is not radically opposed to principles already established; you will have a union of classicism and romanticism.

Art's new departures.

Any poem or painting which produces a serious and lasting impression will in the end be found to have a beauty, not merely of its own, but allied to universal types and susceptible of logical analysis. Its royal stamp will be detected by the expert. Gainsay this, and you count out a host of the elect brotherhood who make this the specific test, — who will forego other elements (as in religion the Church passes over minor matters if you accept its one essential) and concentrate their force upon the dogma tersely expressed by Poe when he defined poetry as "the rhythmical creation of

The æsthetic canon.

beauty." One need not accept this as a sufficient statement, but one may assert that no statement is sufficient which does not pointedly include it.

Confront, however, the fact that the new æsthetic is grounded in science, and see to what this leads. It opposes, for example, the theory of those who accept the existence of a something which we recognize as beauty, and which as a sensible and primary quality can be defined only by itself, or by a synonym, though its conditions are observable and reasons can be given for it. Expression is its source; is not beauty itself, but that which gives objects beauty. Now Véron, a forcible expositor of the school that has in mind the scientific situation, declares that beauty is solely in the eye or mind of the artist, and that everything turns on the expression of his impression. The latter clause is true enough. The beauty which the painter or poet offers us certainly depends upon the quality of his vision, upon his ability to give us something in accord with gen-

Æsthetics as the study of "the manifestations of artistic genius."[1]

Eugène Véron's exposition of "personal art."

---

[1] Véron, in his *L'Esthétique*, declares very justly that the definition of Æsthetics as the "Science of the Beautiful" itself requires defining; that the beauty of Art does not consist in imitation, or realism, or romanticism, but in effects determined by the individuality of the artist; and that herein lies the true worth of impressionism. Finally, he accepts, in deference to usage, the "Science of Beauty in Art" as a convenient formula, but prefers his own statement that "Æsthetics is the science whose object is the study and elucidation of the manifestations of artistic genius."

eral laws, yet deriving a special charm and power from the touch or atmosphere of his personal genius. As each race has its specific mode of vision, so for each there are as many and different impressions and expressions as the race has artists; and the general or academic outlines of perfection being known, the distinctive value of a poem or painting does come from its maker's habit of vision and interpretation.

But why, in order to advance the banner of impressionism, or of neo-impressionism, or of realism, good as these may be, should we assume the task of denying beauty altogether? Beauty is confessedly not a substance; you cannot weigh it with scales or measure it with a yard-stick: but *it lies in a vibratory expression of substances.* It characterizes that substance which enforces upon intelligence — in our case, upon human intelligence — a perception of its fitness. In the mind of a creative poet, it is a quality of his imagined substance, — poetry dealing, as we have seen, with "the shews of things" and treating them as if real. To the pure idealist they are the only realities, as Emerson himself implied in his remark when called away from an abstract discussion in the library to inspect a farmer's load of wood: "Excuse me a moment, my friends. We have to attend to these matters just as if they were real."

Yet Beauty is no less a force existent.

To be sure, from the place where I stand, I cannot see the rays, the vibrations, which convey to you the aspect of something in your line of vision: <small>Its vibrations actually occur,</small> the light and shape and color which constitute your impression are your personal sensations. But the vibrations which produce them are actually occurring, and the quality of the substance from which they emanate is operative, — unless, again, you choose to deny *in toto* the existence of matter, — and, after every allowance has been made for personal variation, if I move to your point of view, they will, so far as we can know anything, produce approximately the same effect upon my mind and upon yours. It matters not through <small>and excite our spiritual perception of them.</small> which of the senses impressions are received: sight, hearing, smell, taste, touch, all resolve themselves at last into spiritual feeling. Form, for example, appeals to the touch as well as to the eye. Note the blind Herreshoff, that skilled designer of the swift, graceful hulls of yachts and other cruisers. As his sensitive fingers passed over, and shaped, and reshaped his model, he had as keen a sense of the beauty of its lines as we have in seeing them. A poem, conveyed by touch to one congenitally deaf, dumb, and blind, will impress him only with the beauty of its thought, construction, and metric concordance; but in one who has lost his sight and hearing in mature years, and who retains his memories, it will excite ideas of <small>Form and sentiment.</small> sound and imagery and color. Moral and intellectual beauty is the spiritual analogue of that which is sensuous; but just now we are regarding concrete qualities; for example, the form, the verbal and rhythmical excellence, of a poet's poem. Our reference to arts that specially appeal to the eye is illustrative, since they afford the diagrams, so to speak, of most service in this discussion.

For the perception of the beautiful there must be a soul in conjunction; that statement is irrefutable. Yet I think that the quality of beauty exists in substances, even if there be no intelligence at hand to receive an impression of it; that if a cataract has been falling and thundering and prismatically sparkling in the heart of a green forest, from time immemorial, and with no human being to wonder at it, it has no less the attribute of beauty; it is waiting, as Kepler said of its Creator, "six thousand years for an interpreter." Suppose that an exquisite ode by Sappho or Catullus has been buried for twenty centuries in some urn or crypt: its beauty is there, and may come to light. Grant that our sense of material beauty is the impression caused by vibrations; then *the quality regulating those vibrations* is what I mean by the "beauty" of the substance whence they emanate. Grant what we term the extension of that substance; the characteristics of that extension are what affect us. There is no escape, you see, unless, with Berkeley, you say there is no matter.

<small>Beauty is the quality regulating certain vibrations.</small>

This is just what the poet, the artist, is not called upon to do. He is at the outset a phenomenalist.

He sets forth his apparitions of things, idealizing them for the delight of himself and the world. And as to the law of beauty, whether it lies in use or proportion or what not, it all comes back to the truths of nature, to the perfection of the universe, to that sense of the fitness of things which is common to us all in our respective degrees; so that there are some objects so perfect that we all, if of the same breed and condition, assent to their beauty. There are women, for example, who take the world with beauty at first glance, and there are other objects only partly beautiful, less perfect, about which, therefore, even critical judgments are in dispute. That beauty does go somewhat with use is plain from its creation by necessity. The vessel that is most beautiful, that differs most from the lines of a junk or scow, is the one best fitted safely and swiftly to ride the waves. The condition is the same with everything in nature and art, from a bird to a portico. If the essence of beauty lies in conformity to the law and fitness of things, then all natural things are as beautiful as they can be,— that is, beauty is their natural quality; they develop it unconsciously as far as possible under limitations imposed by the pervading struggle for existence. That is what leads Hartmann to assert that in Nature's beauty "the individual, who is at the same time marble and sculpture, realizes the Idea perfectly unconsciously; in human artistic production,

The poet receptive to universal phenomena.

Beauty the natural quality of all things.

on the other hand, the instigation of consciousness supervenes."

The poet, through intuition and executive gift realizing the normal beauty of everything, imaginatively sets it forth. He detects it even within the abnormal gloom and deformity imposed by chance and condition, helps it to struggle to the light, restores it — I may say — to consonance with the beauty of the universal soul. This being partly comprehensible through empiricism and logical analysis, men of talent and of little insight produce tolerable work by means of trained æsthetic judgment. But no art, no poetry, is a distinctive addition to the world's stores unless its first conception be intuitive; then only it is a fresh expression of the universal beauty through one of its select interpreters. Like all things else it comes to us from Jove. *Recognition of it is intuitive.*

Even Véron is compelled to assume the element which he denies. When he begins to illustrate and to criticise, he instantly talks of "the perfection of parts." The despotism of established art systems springs from this perfection, — the academic sway of the antique and of Raphaelitism. Much of this discussion belongs to metaphysical æsthetics, and some persons may think these notions antiquated. We know little of these things absolutely. We know not the esoteric truth in matters of art or nature, — otherwise the schools at once would cease their controversies. As it happens, *The schools.*

certain of the latest physicists claim that "deduced facts" — that is, the objects inferred from our sensations — are the true substantialities; that only our perception of them is transient; that the world of subjective feelings is the chimera, not the objective matters which excite perception.

One question you very properly may ask: "Why not take all this for granted, and go on? Join either side, and the result is the same. Eclipses were calculated readily enough upon the Ptolemaic method." Not so. The theory that beauty is a chimera leads to an arrogant contempt for it on the part of many artists and poets, who substitute that which is bizarre and audacious for that which has enduring charm. It begins with irreverence, and leads to discordant taste; to something far beneath the excellence of noble literatures and of great plastic and poetic eras.

<small>Where danger lies.</small>

The tentative revolts that break forth in art and letters are against methods to which, however fine they be and grounded in nature, the world has become too servile. Movements in poetry, like those of Blake and Whitman and Lanier for greater rhythmical freedom, of the Rossettians for a study of Preraphaelite methods, of Banville and Dobson for a restoration of attractive forms; movements in art like those of Monticelli and Claude Monet, — all these are to some extent the quest for values so long unwonted that they seem new; and

<small>"The old order changeth."</small>

thus art returns upon its circuit and the wisdom of the Preacher is reaffirmed. Still, every race has its culminating or concurrent ideal of beauty, which is affected, again, by the conditions of life in the different regions of the race's establishment. Each nation, like a rose-tree, draws from the soil and air its strength, and wealth, and material sustenance; it puts forth branches, and leaves, and sturdy thorns, and battles with the elements and with the thicket that hems it in; finally, with all its hardier growth assured, it breaks into flower, it develops an ideal; its own and perfect rose of beauty marks the culmination, the intent, the absolute fulfilment, of its creative existence. Thus the ideals of Grecian art and song doubtless represent the South, and those of the Gothic or romantic the North, in Europe; and the two include the rarest of our Aryan types. In art, these have resulted in various academic standards the excellence of which cannot be discredited. Pater has rightly said that it is vulgar to ignore the "form" of the one, and vulgar to underrate the "substance" of the other. The charm of the antique, for instance, is so celestial that, supposing we had been deprived of it hitherto and were suddenly to be introduced to it through discovery of a new continent, the children of art would go wild over its perfection. The very artists who now revolt from it would in that case break from other standards and lead a revolt in its favor, and a

Development of racial and national ideals.

Ultimate standards.

Perfection of the antique.

momentous progress in art and song would be recorded. As it is, we are intellectually aware of its nobility; but anon our sense of delight in it is blunted, — we have no zest in its repetition, being to the manner born. Zest is the sensation most worth possessing. The eager student instinct is right in essaying discovery and revival, since only thus can zest be sustained, and, for the sake of this, occasional changes even to fashions of minor worth are not to be scouted. The element of strangeness itself conveys a peculiar effect of beauty. This, by the way, is the strength of the Grotesque, a subordinate form of art and at its best accessory.

<small>Fashion. Cp. "Poets of America": pp. 273, 274.</small>

You will observe that after most revolts the schools go back, in time, to certain ideals, — to those which become academic because the highest. They recover zest for these, having wearied of some passing fashion or revival. An occasional separation is not a bad thing, after all, in friendship, art, or marriage. Thus it was that the classic Renaissance of Italy reopened a world of beauty, and began a fresh creative period, in which new styles of painting, moulding, architecture arose, different from the antique, but inspired by it, and possible because the spirit of beauty itself was reborn.

<small>Renaissance.</small>

We constantly have illustrations of the dependence of artistic zest upon the stimulus of novelty. Some of you possibly were brought up in our old towns and in those old houses where architecture, furniture, wall-paper, were all "in keeping." How prim and monotonous it then seemed, and how a lad longed to get away from it! Citified folk long since got away, and with zest, to something vastly inferior,—to something with no style at all. At last the Colonial and Revolutionary homestead styles became rare to find in their integrity. Now we see a restoration of them; now we rediscover their lightness and fitness,— their beauty,—and are reviving them in all departments of taste; until, in fact, as I recently heard an artist break forth, "there is a great deal of taste,— and some of it is good!" It may be that another generation will tire of them, as we did, though it seems heresy to say so now. *Another illustration: the "Colonial" revival.*

For a long time after 1775, Sir Joshua Reynolds stood, in his work and "Discourses," as a representative exponent of the academic. One must remember that he had no light task in promoting taste among his Anglo-Saxons; their race is not endowed with the intuitive Southern perception of the beautiful. The English acquire their artistic taste intellectually, except in landscape-gardening, although their poets seem to be even more noble (perhaps because more intellectual) than those of nations whose sense of material beauty is congenital. Sir Joshua was a good deal of a poet with his brush. The chief of academi- *Academic art.* *Sir Joshua Reynolds.*

cians, he had a touch, a lovely feeling, an impressiveness of his own. When he sought a foundation for his discourses upon art, he wisely went to the best ideals known to him. His lectures are in the main sound; no artist, even a recanter, can afford not to read them; yet the attempt to carry them out almost confirmed the English School in "correct," rigid, and lifeless methods. And why? Because Sir Joshua, an original painter in his studio, in his teachings did not sufficiently allow for and inculcate a local, climatic, racial divergence from his revered Italian models.

Now, the Indo-European ideals of beauty usually have been the foundation of academic theoretics upon art, just as they are interwrought, in sooth, with English poetry, and with the great criticism thereon, — from Lamb and Coleridge to Dryden and Arnold and Lowell. But what would Sir Joshua Reynolds have made of the extreme antipodal type, that of those Asiatic Greeks, — our delightful Japanese? To be sure, there were Indian and Chinese cults, but these were merely capricious and accessory, and not pursued to any just appreciation of their ideals. Here, then, in Japan is a race developed under distinctive biological conditions, with types of art and life almost the reverse of our own, yet perfectly consistent throughout, and — as we now see — superior to those of Western civilization in more than one department. Its ideals are just as perfect as those of the Greeks or Goths, yet absolutely different. Here we indeed enter a new world. Ideal beauty plainly lies in adaptation of the spirit to the circumstances, though not always to the apparent material exigencies. [*Fitness, material and spiritual.*] La Farge, whom I have before quoted, — and upon the subject of beauty the sayings of a painter or an architect (*mutatis mutandis*) apply just as fully to poetry as to his own art, — La Farge says, in speaking of the adaptation of Japanese buildings to resistance against earthquakes, that

"like all true art, the architecture of Japan has found in the necessities imposed upon it the motives for realizing beauty, and has adorned the means by which it has conquered the difficulties to be surmounted."

No better illustration could be given of the relations of fitness and beauty; but he soon has occasion to add: —

"Everywhere the higher architecture, embodied in shrines and temples, is based on some ideal needs, and not essentially upon necessities."

We see, then, every people recognizing an extramundane conception of beauty, founded in the spirit of man, and this again conforms itself to the spirit of each race. Through it the poets become creative rather than adaptive, — the beauty of their imaginings coming from within, just as the beauty of nature is the efflux of the universal spirit. So far as human artists share the Divinity of that spirit, their interpretations give it form to human eyes, melody to human ears, and imagery and feeling therewithal to move the recipient. It seems, then, I say, the lot of each nation, as if an individual, and of each period, as if a modish season, to discover the beauty conformed both to general laws and to specific needs and impulse; to create, moreover, its proper forms in every art, thus making new contributions to the world's thesaurus of poetry and design. This is acknowledged by all, as concerns the every-day art of dress. A Japanese gentleman is dignified in his national costume; his wife and daughters are charming in their clinging and curving robes. Attire them — and that is the shameful thing which our invasion is effecting — in the dapper broadcloth, the Parisian gown, and their comeliness often is gone. A pitiful incongruity is apt to take its place. I believe that such a race as theirs also develops its fine arts, manners, government, literature, — yes, even religion, — to its foreordained capacity; that if you force or coax it to adopt the modes of a divergent people, you sound the death-knell of its fair individuality. If the tempter race is the superior, the one that surrenders its own ideals is doomed to be absorbed, — at least, to lose its national distinction. Possibly with the progressive modern intercourse of peoples a general blending is to result. Languages, arts, races, may react upon one another and produce

Specific evolution.

What distant goal?

a cosmic mongrelism. If this is according to the law of progress, something grand will come out of it, a planetary and imposing style. But during centuries of transition the gradual loss of national individualities will seem pathetic indeed. Something of this passed through my mind as I watched, half sorrowful and half amused, an accomplished Japanese lady, the adopted daughter of an American, yielding to the influence of our Western ideals. A natural artist, like so many of her blood, she is impressible by beauty of a novel type. As far as personal experience is concerned, she doubtless adds to the worth of her own life by assimilating the results of an art no more perfect in its kind than the decorative — and therefore secondary — art of her own race, yet one far beyond the power of her race to originate, or to pursue in competition with its originators. Therefore it seemed almost a pity to find her at work upon a lesson from the Art Students' League, copying in crayon an antique Apollo, with deft fingers, which to my thinking should be tracing designs in lacquer or in cloisonné on bronze, or painting some group of Japanese men and maidens, in their flexible costume, by the bayside, on a terrace, with herons stalking among sacred lilies in the near distance, and the eternal peak of Fujiyama meeting the blue sky beyond. *The assimilative process.*

Meanwhile our present standard of beauty is the European, with modifications. To comprehend any other you must enter into its spirit by adoption, by a certain naturalization; until then you will find it as hard to master as the idioms of a language not your own. These seem grotesque and childish until you speak, even think, in their tongue without mentally translating it. A translation will give you the imagination, action, thought, of a poem, for instance, but not its native and essential *beauty*. Æsthetics relate to the primal sense, and must be taken at first hand. This is all the truth there is in the maxim *De gustibus*. If the rays of our sun were as green as those of the star β Libræ, beauty would exist and have its standard in conformity. Taste would be as intuitive as now, and just as open to cultivation.

<small>Taste is congenital yet cultivable.</small>

These general principles should entitle us to our surmise respecting the ultimate value of a poem. A mode attractive for its novelty may be only the vogue of a generation, or of a brief season. I take endurance to be the test of art. History will show, I think, that if a poem had not the element of beauty, this potency in art, its force could not endure. Beauty partakes of eternal youth and conveys its own immortality. Passion and imagination intensify much of the poetry that has survived; but under their stress the poet summons beauty to his aid. Wisdom and morals do not so inevitably take on grace; their statements, impressive at the time, must be recast perpetually. The law of natural selection conserves artistic beauty

<small>Poetic beauty.</small>

<small>Its conserving power.</small>

in the poem as in the bird and butterfly. Besides, just as gems and gold are hoarded while iron is left to rust, and as paintings that are beautiful in line and color grow costlier with time, so the poetry that has the beauty of true art becomes the heirloom of generations. For beauty seems to consecrate both makers and possessors. Just as all the world clings to the legends of Helen and Cleopatra and Mary Stuart, so it has a fondness for the Cellinis and Villons and Marlowes and Lovelaces,— the ne'er-do-weels of art and song. This is because it reads the artist's higher self in his work; there alone it is expressed, and we give him credit for it. The truth of fairy tales is that of beauty; the Florizels and Cinderellas and Percinets are its ideals. Beauty loves the Beast, but the Beast is beauty in disguise. Thus creative taste holds the key to the future, and art for art's sake is a sound motto in so far as beauty is a legitimate end of art. That it is not the sole end of art-life is the lesson of Tennyson's "The Palace of Art." One who thought otherwise at last found need to throw her royal robes away: _{Art for art's sake.}

> "Make me a cottage in the vale," she said,
> "Where I may mourn and pray.
>
> "Yet pull not down my palace towers, that are
> So lightly, beautifully built:
> Perchance I may return with others there
> When I have purged my guilt."

All in all, if concrete beauty is not the greatest

thing in poetry, it is the one thing indispensable, and therefore we give it earliest consideration. Besides, it so depends on the elements of emotion and truth that when these are not expressed in a poem you may suspect the beauty to be defective and your sense of it mistaken. It may be said to symbolize truth in pure form.

<small>The poet's instinct.</small> The young poet, as instinctively as a plant seeks the light, feels that he must worship and express the beautiful. His passion for it, both in his life and in his art, is his greatest strength and danger. It is that which must distinguish him from other men; for many will have more wisdom, more virtue, than himself, while only he who can inform these with beauty by that token is the poet. In the early poems of Shakespeare, Milton, Goethe, Tennyson, Rossetti, thought is wreaked "upon expression." Even the lyrists whose development stops at this point, such as Herrick, the cavalier singers, the Provençal minstrels, have no obscure stations in the hemicycle of song.

<small>Survival of the fittest and most beautiful.</small> Why is it that all the relics of Grecian poetry have such beauty? Were there no dullards, was there no inartistic versifying, even in Athens? It is my belief that for every poet whose works have reached us a score passed into obscurity, and their writings were lost; furthermore, that, in spite of the burning of the Alexandrian library, comparatively little has been lost since the

time of Herodotus that was worth saving. Only the masterpieces, large and small, were copied and recopied, and treasured in men's hearts and homes. And those were. The ugly statues, also, went to ruin. It is the Venus of the Louvre that is piously buried when danger threatens, whether in Melos or by the Seine; and it is she who always rises again and comes to light. Doubtless we have the *most* beautiful dramas of even Æschylus and Sophocles, and some of the choicest verse of even Æolian and Dorian lyrists. Belief in this is not shaken by the recovery of classical fragments through our archæological explorations; for, if something fresh and fair — a portion of the Antiope, for instance — is occasionally gained, it is surprising how many passages from works already in our hands are quoted in the writings upon new-found tablets and papyri. Time and fate could not destroy the blooms of the anthology, the loveliest Syracusan idyls, the odes of Catullus and Horace. By chance something less attractive has remained: we keep Ausonius and Quintus on the archaic shelves, but they have no life; they are not cherished and quoted, they cannot be said to endure.

All service is in a sense acceptable, and hence the claim that the intent, rather than the outcome, crowns the work. Thus Browning in his paper on Shelley and in certain poems shows himself to be a pure idealist in his estimate of art. Professor A. H. Smyth explains that the object of

Motive and accomplishment.

Browning's "Old Pictures in Florence" is "to show that Greek art in all its matchless perfection is no more admirable than dim and almost undecipherable ruins of efforts merely monastic, on smoke-stained walls of Christian churches." But to me the latter suggest merely faith and aspiration, without that perfected beauty which adds the grandeur of attainment and completes the trinity of art.

The poetry of our own tongue is sufficient to test the law of durability. Its youth, as if that of a poet, was pledged to the mastery of the beautiful as soon as it grew out of half-barbaric minstrelsy and displayed a conscious intent. Chaucer is a poet of the beautiful; always original in his genius, and sometimes in his invention, he for the most part simply tells old tales with a new and English beauty. Five hundred years later his pupil, Morris, renews the process. Spenser's rare and exhaustless art makes him a poet for poets. Passing by Shakespeare as we would pass by nature, what we cull again and again from the Elizabethan garden are those passages in the dramatists, beautiful for rhythm and diction, which furnish examples for the criticism of Coleridge and Lamb. From the skylark melodies and madrigals of that English Arcady those which are most beautiful are ever chosen first by the anthologists. We never tire of them: they seem more perfect and welcome with each remove. Too few read Ben Jonson's plays; who does not know "To Celia," "The Triumph of

Charis," and "Drink to me only with thine eyes"? The song, "Take, O take those lips away," even were it not embalmed by Shakespeare, would outlast the dramas of John Flêtcher. Suckling's "Why so pale and wan, fond lover?" and his verses on a wedding; Lovelace's "To Lucasta" and "To Althæa, from Prison," — such are the gems in whose light the shades of courtier-poets remain apparent. More of Herrick's endure, because with him beauty of sound and shape and fancy was always first in heart, and always fresh and natural. I have written a paper on Single-Poem Poets, but the greater number of them were no less the authors of a mass of long-forgotten verse. Of Waller's poetry we remember little beyond the dainty lyrics, "Go, lovely rose" and "On a Lady's Girdle." From time to time the saddest and gladdest and sweetest chansons of Villon and Ronsard and Du Bellay are retranslated by deft English minstrels, as men take out precious things from cabinets and burnish them anew. A ponderous epic disappears; some little song, once carolled by Mary Stuart, or a perfect conceit of imagery and feeling, whose very author is unknown, becomes imperishable. For instance,

THE WHITE ROSE.

Sent by a Yorkish Lover to his Lancastrian Mistress.

"If this fair rose offend thy sight,
Placed in thy bosom bare,

'T will blush to find itself less white,
And turn Lancastrian there.

" But if thy ruby lip it spy,
As kiss it thou mayest deign,
With envy pale 't will lose its dye,
And Yorkish turn again."

The few lyrics I have named are among the most familiar that occur to you and me; but what has made them so if it be not their exceeding loveliness?

We have but one poet of the first order, but one *From Shakespeare to Wordsworth.* strong pier of the bridge, between Shakespeare and our own century. Milton in his early verse, which has given lessons to Keats and Tennyson, displays the extreme sense and expression of poetic beauty. Dryden and Pope have values of their own; but from Pope to Burns, only Goldsmith, for his charms of simplicity and feeling, and Collins and Gray, who achieved a certain perfection even in conventional forms, are still endeared to us. Examine the imposing mass of Wordsworth's poetry. With few exceptions the imaginative and elevated passages, the most tender lyrics, have a peculiar beauty of rhythm and language, — have sound, color, and artistic grace. Take these, and nearly all are chosen for Arnold's "Selection" and Palgrave's "Golden Treasury," and you possibly have the most of Wordsworth that will be read hereafter.

A revival of love for the beautiful culminated in

the modern art school. Naturalness had come back with Burns, Cowper, and Wordsworth; intensity and freedom with Byron; then the absolute poetic movement of Coleridge, Shelley, Keats, and of that æsthetic propagandist, Leigh Hunt, began its prolonged influence. Poetry is again an art, constructed and bedecked with precision. So potent the charm of this restoration, that it has outrun all else: there is a multitude of minor artists, each of whom, if he cannot read the heart of Poesy, casts his little flower beside her as she sleeps. Who can tell but some of these blossoms may be selected by Fame and Time, that wait upon her? Ars Victrix wears her little trophies as proudly as her great. Dobson's paraphrase on Gautier became at once a proverb, from instant recognition of its truth:—

_{Modern æstheticism.}

_{"L'art robuste seul a l'éternité."}

> "All passes. Art alone
>   Enduring stays to us;
> The Bust outlasts the throne,—
>   The Coin, Tiberius;
>
> Even the gods must go;
>   Only the lofty Rhyme
> Not countless years o'erthrow,—
>   Not long array of time."

In this one lecture, you see, I dwell upon the technical features that lend enchantment to poetry in the concrete. How, then, does the beauty of a poem avail? Primitively, as

_{Elements of concrete poetic beauty.}

addressed to the ear in sound; that was its normal method of conveying its imagery and passion to the human mind, and we have already considered the strange spell of its vocal music. But with the birth of written literature it equally addressed the eye, and since the invention of printing, a thousand times more frequently; so that the epigram is not strained which declares that "It is read with the ear; it is written with the voice; it is heard with the eyes." The mind's ear conceives the beauty of those seen but "unheard melodies" which are "the sweetest." The *look* of certain words conveys certain ideas to the mind; they seem as entities to display the absolute color, form, expression, associated with their meanings, just as their *seen* rhythm and melody sound themselves to the ear. The eye, moreover, finds the architecture of verse effective, realizing a monumental, inscriptionary beauty in stanzaic and ode forms. Shape, arrangement, proportion compose the synthetic beauty of Construction. Thus poetry has its architecture and shares that condition celebrated by Beatrice in the "Paradiso": "All things collectively have an order among themselves, and this is form, which makes the universe resemble God."[1] Beauty of construction is still more potent in the effect of plot and arrangement. Simplicity, above all, characterizes alike the

[1] Thus cited by Dr. W. T. Harris in *The Spiritual Sense of Dante's Divina Commedia*.

noblest and the loveliest poems, — simplicity of art
and of feeling. There are no better ex- Simplicity.
amples of this, as to motive and construction, than
those two episodes of Ruth and Esther. Written in
the poetic Hebrew, though not in verse, Examples, and
they fulfil every requisition of the prose a contrast.
idyl: the one a pure pastoral, the other a civic and
royal idyl of the court of a mighty king. There is
not a phrase, an image, an incident, too much or too
little in either; not a false note of atmosphere or
feeling. These works, so naïvely exquisite, are
deathless. Their charm is even greater as time
goes on. Now, a remarkable novel has been written
in our own day, "Anna Karénina," which chances
to be composed of two idyls,— one distinctly of the
city and the court, the other of the country and the
harvest-field. These two cross and interweave, and
blend and separate, until the climacteric tragedy and
lesson of the book. Powerful as this work is, it has
little chance of great endurance, inasmuch as its
structure and detail are complex even for this complex period. It is at the opposite extreme from the
simplicity of those matchless idyls of the Old Testament.

Nevertheless, that idyllic perfection came from a
really advanced art. However spontane- Nature of
ous of impulse, it was not perpetuated simplicity.
through the uncertain process of oral transmission,
but by a polished scriptural text. Absolutely primitive song was often a rhapsody, and not suited to

textual embodiment. When finally gathered up from traditions, it owed as much to the compiler as a rude folk-melody owes to a composer who makes it a theme for his sustained work. The judge of such poetry, then, must consider it as both an art and an impulse, and even as addressed to both the eye and the ear. And while it is true that the simplicity of the ancients, of purely objective art, is of the greatest worth, we must remember that the works in question were the product of an age of few "values,"

<small>Our compensation for its loss.</small> — as a painter would say. In our passage from the homogeneous to the complex, the loss in simplicity is made up by the gain in variety and richness. We return to simplicity, ever and anon, for repose, and for a new initiative, as a sonata returns to its theme. Refreshed, we advance again, to still richer and more complex inventions. In place of the few Homeric colors we have captured a hundred intermediate shades of the spectrum, and we possess a thousand words to recall these to the imagination. The same progression affects all the arts. What modern painter would be content with the few Pompeian tints; what musician with the five sounds of the classic pentachord?

Artistic simplicity, then, must be attained through <small>The natural key.</small> naturalness; and from that grace of graces modern complexity of material and emotion cannot debar us. If a poet, imitating antique or foreign methods, confines himself baldly to a few "values," he may incur the charge of artifice; and artificiality is the antithesis of naturalness. You may exhibit an apparent simplicity of style and diction, — which Mrs. Browning, for instance, failed of altogether, — and yet have no sincere motive and impulse, in respect of which her lyrics and sonnets were beyond demur.

In poetry true beauty of detail is next to that of construction, but non-creative writers lavish all their ingenuity upon decoration until it becomes a vice. You cannot long disguise a lack of native vigor by ornament and novel effects. Over-decoration of late is the symptom of over-prolonged devotion to the technical sides of both poetry and art. Sound, color, word-painting, verse-carving, imagery, — all these are rightly subordinate to the passion of a poem, and must not usurp its place. Landscape, moreover, at its best, is but a background to life and action. In fine, construction must be decorated, but decoration is not the main object of a building or a poem. "The Eve of St. Agnes" is perhaps our finest English example of the extreme point to which effects of detail can be carried in a romantic poem. The faultless construction warrants it. Some of Tennyson's early pieces, such as the classico-romantic "Œnone" and "The Lotos-Eaters," stand next in modern verse. But I forego a disquisition upon technique. All of its countless effects are nothing without that psychical beauty imparted by the true poetic vitality, — are of

<small>Beauty of detail.</small>

<small>Over-elaboration. Cp. "Victorian Poets": p. 289.</small>

<small>One thing needful.</small>

less value than faith and works without love. The *vox humana* must be heard. That alone can give quality to a poem; the most refined and artistic verse is cold and forceless without it. A soulless poem is a stained-glass window with the light shining on and not through it.

Since a high emotion cannot be sustained too long without changing from a rapture to a pang, many have declared that the phrase "a long poem" is a misnomer. Undoubtedly, concentration· of feeling must be followed by depression or repose. The fire that burns fiercely soon does its work. Yet he who conceives and makes a grand tragedy or epic so relieves his work with interludes and routine that the reader moves as from wave to wave across a great water. It may be, as alleged, a succession of short poems, but these are interwrought as by one of nature's processes for the building of a master-work. However, let me select the beauty of a short and lyric poem, as the kind about which there is no dispute, for the only type which I can here consider.

<small>Mill's canon, afterwards sustained by Poe.</small>

Lyrical beauty does not necessarily depend upon the obvious repetends and singing-bars of a song or regular lyric. The purest lyrics are not of course songs; the stanzaic effect, the use of open vowel sounds, and other matters instinctive with song-makers, need not characterize them. What they must have is *quality*. That their rhythmic and verbal expression appeals supremely to the

<small>Lyrical beauty:</small>

finest sensibilities indicates, first, that the music of speech is more advanced, because more subtly varying, than that of song; or, secondly, that a more advanced music, such as the German and French melodists now wed to words, is required for the interpretation of the most poetic and qualitative lyric. A profound philosophy of sound and speech is here involved,—not yet fully understood, and into which we need not enter.

But you know that rare poetic types, whether of the chiselled classic verse, or of the song and lyric, have a grace that is intangible. *its subtile quality.* There is a rare bit of nature in "The Reapers" of Theocritus. Battus compares the feet of his mistress to carven ivory, her voice is drowsy sweet, "but her air,"—he says,—"I cannot express it!" And thus the gems of Greek and Latin verse, the cameos of Landor and Hunt and Gautier, the English songs from Shakespeare to Procter and Tennyson and Stoddard, the love-songs of Goethe and his successors, the ethereal witching lyrics of Shelley and Swinburne and Robert Bridges,—all these have one impalpable attribute, light as thistle-down, potent as the breath of a spirit, a divine gift unattainable by will or study, and this is, in one word, Charm. Charis, Grace herself, bestows *Charm.* it, blending perfect though inexplicable beauty of thought with perfect though often suggested beauty of feeling. To these her airy sprites minister with melody and fragrance, with unexpectedness and sweet surprises, freedom in and out of law, naïveté, aristocratic poise, lightness, pathos, rapture, — all gifts that serve to consecrate the magic touch. However skilled the singer, quality and charm are inborn. Something of them, therefore, always graces the folk-songs of a peasantry, the ballads and songs, let us say, of Ireland and Scotland. Theirs is the wilding flavor which Lowell detects: —

> "Sometimes it is
> A leafless wilding shivering by the wall;
> But I have known when winter barberries
> Pricked the effeminate palate with surprise
> Of savor whose mere harshness seemed divine."

When to this the artist-touch is added, then the wandering, uncapturable movement of the pure lyric — more beautiful for its breaks and studied accidentals and most effective discords — is ravishing indeed: at last you have the poet's poetry that is supernal. Its pervading quintessence is like the sheen of flame upon a glaze in earth or metal. Form, color, sound, unite and in some mysterious way become lambent with delicate or impassioned meaning. Here beauty is most intense. Charm is the expression of its expression, the measureless under-vibration, the thrill within the thrill. We catch from its suggestion the very impulse of the lyrist; we are given the human tone, the light of the eye, the play of feature, — all, in fine, which shows the poet in the poem and makes it his and not another's.

The pure lyric. — "Das Durchcomponirt."

Just as this elusive beauty prevails, the song, or lyric, will endure. Art is in truth the victress when she fulfils Ruskin's demand and is able "to stay what is fleeting, and to enlighten what is incomprehensible; to incorporate the things that have no measure, and immortalize the things that have no duration." And yet, recognizing her subtle paradoxy, and if asked to name one suggested feeling which more than others seems allied with Charm and likely to perpetuate its expression (for I can name only one to-day), I select that which dwells not upon continuance, but upon — our perishableness. Think of it, and you will see that Evanescence is an unfailing source of charm. Something exquisite attaches to our sense of it. The appeal which a delicate and fragile thing of beauty makes to us depends as much upon its peril as upon its rarity. In the fulness of life we may have other things as fair and cherished; but that one individuality, that grace and sweetness, cannot be repeated. In time we must say of it:— *[sidenote: Art's beauteous paradox.]* *[sidenote: Most fair because most fleeting.]*

> "Like the dew on the mountain,
> Like the foam on the river,
> Like the bubble on the fountain,
> Thou art gone, and forever!"

We marvel at the indestructible gem, but love the flower for its share in our own doom. If the violet, the rose-gerardia, the yellow jasmine, were unfading, imperishable, what would their worth be? Mimic them exactly in wax, reproduce even their fragrance, and the copies smack of embalmment. We have, indeed, blooms that do not wither, that do not waste themselves in exhalations; we call them immortelles, but we feel that these amaranthine, husky blossoms are emblems not of life but of death; they cannot have souls, else they would not be so changeless. Not theirs

> "The unquiet spirit of a flower
> That hath too brief an hour."

The ecstatic charm of nature lies in her evanishments. Each season is too fair to last; no sunrise stays; "the rainbow comes and goes;" the clouds change and fleet and fade to nothingness. Thus sadness dwells with beauty, —

> " Beauty that must die;
> And Joy, whose hand is ever at his lips
> Bidding adieu."

The height of wisdom, then, is to make the most of life's best moments, to realize that "it is their evanescence makes them fair." So it is with all mortal existence: we idealize the unalterable fact of its mortality. Time passes like a bird, joy withers, even Love dies, and the Graces ring us to his burial. We ask, with the Hindu Prince, concerning life, —

Morituri salutamus.

> "Shall it pass as a camp that is struck, as a tent that is gathered and gone
> From the sands that were lamp-lit at eve, and at morning are level and lone?"

We ask with sighs and tears, but would we have it otherwise? If Poe was wrong in restricting poetry to the voices of sorrow and regret, he was right, methinks, in feeling these to be among the most effectual of lyrical values. The word *Irreparable* suggests a yearning as infinite as that for the Unattainable, under the spell of which Richter fled as from a passion too intense to bear. Yes; the sweetest sound in music is "a dying fall." "Mimnermus in Church" weighs the preacher's adjuration, and makes an impetuous reply:— From Cory's "Ionica."

> "Forsooth the present we must give
> To that which cannot pass away!
> All beauteous things for which we live
> By laws of time and space decay.
> But oh, the very reason why
> I clasp them is because they die."

Among priceless lyrics from the Greek anthology to our own, those of joy and happy love and hope are fair indeed, but those which haunt the memory turn upon the escape — not the retention — of that which is "rich and strange." Their charm is poignant, yet ineffable. The consecration of such enduring melody to regret for the beloved, whose swift, inexplicable transits leave us dreaming of all they might have been, is the voice of our desire that their work, even though perfecting in some unknown region, may not wholly fail upon earth, — that their death may not be quite untimely.

How subtile the effect, even in its English ren-

dering, of Villon's "Ballade of Dead Ladies"—
"Where are the snows of yester-year?"
Are any lyrics more captivating than our
English dirges,—the song dirges of the dramatists:
"Come away, come away, Death," "Call for the
robin redbreast and the wren," "Full fathom five
thy father lies," and the like? Collins' "Dirge for
Fidele," a mere piece of studied art, acquires its
beauty from a flawless treatment of the master-
theme. Add to such art the force of a profound
emotion, and you have Wordsworth in his more
impassioned lyrical strains: "She dwelt among the
untrodden ways," "A slumber did my spirit steal;"
and the stanzas on Ettrick's "poet dead." Landor's
"Rose Aylmer" owes its spell to a consummate
union of nature and art in recognition of the una-
vailability of all that is rarest and most lustrous:—

The ecstasy of pathos.

> "Ah, what avails the sceptred race!
> Ah, what the form divine!
> What every virtue, every grace!
> Rose Aylmer, all were thine.
> Rose Aylmer, whom these wakeful eyes
> May weep, but never see,
> A night of memories and of sighs
> I consecrate to thee."

—Of memories and of sighs, yet not of pain, for
such vigils have a rapture of their own. The perished have at least the gift of immortal love, remem-
brance, tears; and at our festivals the unseen guests
are most apparent. Thus the tuneful plaint of sorrow, the tears "wild with all regret," the touch that

consecrates, the preciousness of that which lives but in memory and echo and dreams, move the purest spirit of poesy to sweep the perfect minstrel lute. To such a poet as Robert Bridges the note of evanescence is indeed the note of charm, and in choosing the symbols of it for the imagery of his most ravishing song,[1] he knows that thus, and thus most surely, it shall haunt us with its immortality:— *"My song be like an air!"*

> "I have loved flowers that fade,
>   Within whose magic tents
> Rich hues have marriage made
>   With sweet unmemoried scents—
> A honeymoon delight—
> A joy of love at sight,
> That ages in an hour:—
> My song be like a flower!
>
> "I have loved airs that die
>   Before their charm is writ
> Upon a liquid sky
>   Trembling to welcome it.
> Notes that, with pulse of fire,
> Proclaim the spirit's desire,
> Then die and are nowhere:—
> My song be like an air!
>
> "Die, song, die like a breath
>   And wither as a bloom:
> Fear not a flowery death,
>   Dread not an airy tomb!
> Fly with delight, fly hence!
> 'T was thine love's tender sense
> To feast, now on thy bier
> Beauty shall shed a tear."

[1] *Poems by Robert Bridges.* Oxford, 1884.

VI.

TRUTH.

IF all natural things make for beauty, — if the statement is well founded that they are as beautiful as they can be under their conditions, — then truth and beauty, in the last reduction, are equivalent terms, and beauty is the unveiled shining countenance of truth. But a given truth, to be beautiful, must be complete. Tennyson's line, *[What is meant by the Unity of Beauty and Truth.]*

> "A lie which is half a truth is ever the blackest of lies,"

will bear inversion. Truth which is half a lie is intolerable. A certain kind of preachment, antipathetic to the spirit of poesy, has received the name of didacticism. Instinct tells us that it is a heresy in any form of art. Yet many persons, after being assured by Keats that the unity of beauty and truth is all we know or need to know, are perplexed to find sententious statements of undisputed facts so commonplace and odious. Note, meanwhile, that Keats' assertion illustrates itself by injuring the otherwise perfect poem which contains it. So obtrusive a moral lessens the effect of the "Ode on a Grecian Urn." In other words, the *[The didactic heresy.]*

beauty of the poem would be truer without it. Now, why does a bit of didacticism take the life out of song, and didactic verse proclaim its maker a proser and not a poet? Because pedagogic formulas of truth do not convey its essence. They preach, as I have said elsewhere, the gospel of half-truths, uttered by those who have not the insight to perceive the soul of truth, the expression of which is always beauty. This soul is found in the relations of things to the universal, and its correct expression *is* beautiful and inspiring.

<small>Half-truths are odious.</small>

While the beautiful expresses all these relations, the didactic at the best is the expression of one or more of them, — often of arbitrary and temporal, not of essential and infinite, relations. We therefore detest didactic verse, because, though made by well-intentioned people, it is tediously incomplete and false.

Poets will interpret nature truthfully, within their liberties; they do not assume to be on as close terms with her, or with her Creator, as some of the teachers and preachers. They are content to find the grass yet bent where she has passed, the bough still swaying which she brushed against. They feel that

"What Nature for her poets hides,
'T is wiser to divine than clutch."

The imaginative poets, who read without effort the truth of things, have been more faithful in even their passing transcripts of nature and life than many who conscientiously attempt a portrayal. Where they make comments, it is as if by anticipation of the reader; it is not so much their own conclusion as that of the observing world. The truth, moreover, is less in the comment than in the poetry, — is rather in the song than in the obligato. With the epic or dramatic poet the motive is not truth of description, but truth of life. Yet how much surer the scenic touches of the best narrative and drama than the word-painting of the so-called descriptive poets! Compare the sudden landscape, the life of its populous under-world, the sky and water, the sunlight and moonlight and storm, in "A Winter's Tale" and "Midsummer Night's Dream," with the prolonged and pious descriptions in Thomson's "Seasons." In the dramas the scenic truth is incidental, yet almost incomparable for beauty; in the descriptive poem it is elaborate and tame. You are comparing, to be sure, the greatest of poets with one relatively humble, but the latter is on his chosen ground, and gives his whole mind to his business. Something more than sincerity and knowledge, then, is needed for the expression of truth. Superadd noble contemplation and the anointed vision that reads the life of nature, and you have Wordsworth, a poet and painter indeed. In his greater moods he assuredly sets us face to face with unadulterate truth. Even Wordsworth does this

Truth a matter of course in the best art.

Faculty better than intention.

less effectively, when his interpretation is premeditated, than certain bards whose side-glimpses of the outdoor world we interpret for ourselves. Their chance strokes are matchless. The classic isles and waters are all before us in the "Odyssey," characterized broadly and truthfully by essential traits. Attica glows and glooms in the choruses of "Œdipus at Colonos" and "The Clouds;" we have the atmosphere that suffuses her landscape, action, personages. Its tone is just as capturable now as two thousand years ago under the sky of Sophocles and Aristophanes. The phonograph passes no more intelligibly to after time the living voice of a Gladstone or a Browning. Rarely is there an avowedly descriptive poet who achieves much more than the asking you to take his word for a mass of details. To come near home, this was what such American landscapists as Street and Percival usually succeeded in doing; while Lowell, with his quick eye and Greek good-fellowship with nature, always keeps us in mind of her as a blithe companion by his side when he chats to us, and whether on the rocks of Appledore, or under the willows, or along the snow-paths of a white New England night. Cowper got nearer to truth than Thomson; he pointed to the naturalness that Wordsworth sought in turn, — and found. As for Burns, he lay in nature's heart, and — whether with or without design — expressed her as simply and surely as the bards of old.

> "Descriptive" poets. Cp. "Poets of America": p. 46.

Of both truth to life and truth to physical nature there are two poetic exhibits: the first, broad; the second, minute and analytic. *Breadth, and universality.* The greater the poet, the simpler and larger his statement, however fine in detail when need be. Seeking that presentment of human character and experience which is universal, we go to the poets and idylists of the Bible, to Homer and the Attic dramatists, to Cervantes and Shakespeare, to Molière, and to the great novelists of the modern age. In poetry life has never been treated at once with so much intensity and truth, by many contemporaries, as in the Elizabethan period. This was inevitable. Our early dramatists *The Elizabethans.* wrote for instant stage production; their poetic text was of much import in default of the perfected acting and accessories which now render the text less essential, — in fact, far too subordinate. In such "effects" as the stage production then made practicable, Shakespeare and his group have not been excelled. But life — truth of life and character — then was all in all; a false transcript was instantly detected; the dramatic poet, however exuberant, founded his work in unflinching realism. Situations and trivial sentiment now make the playwright, and even Tennyson and Browning have been unable to restore the muse conspicuously to the stage. The laureate's genius, to be sure, is the reverse of dramatic. Browning had the requisite passion and dramatic instinct; life and motive engrossed him beyond all else. But contrast the bold, direct Elizabethan characters with Browning's personages,—whose thought and action are analyzed by him to the remotest detail. His drama is unique, but not in the free and instant spirit of poetry; it is not so much life as biology. The distinction recalls that tradition of the Massachusetts bar. Webster and Choate often were opposed in leading cases. The former brought his power and learning to bear upon the main issue of a case, and brushed aside the inessentials. Choate delighted to follow every trail to the uttermost, and in a manner as analytic as that of "The Ring and the Book." The jurors marvelled at Choate's intellectual dexterity and glitter, but Webster usually won the verdict. The jury of an author is the reading world. In prose romance America puts forward a counterpart to Browning,—Mr. Henry James, except that he never sacrifices an imperturbable refinement of style; besides, with reference to his novels at least, he usually avoids, as if on principle, the concentrated passion and the dramatic situations that at times make Browning so impressive.

<small>The analytic method. Cp. "Victorian Poets"; p. 432.</small>

On the other hand, when Browning, the anatomist of human life, interests himself with side-glimpses of nature, he is full of simple truth, and with a sure instinct for essentials. His lyrics abound in these beautiful surprises. He forgets the laboratory when he touches landscape and outdoor life, and is all the artist. Nature has but

<small>Browning, Tennyson, etc.</small>

one truer painter among the dramatists, and the best touches of both seem incidental. When Browning thinks of birds and beasts they suddenly, as in the Arabian Nights, become almost human. He reads the heart, one might say, of a bird, a horse, or a dog. This Tennyson does not do, nor does he usually give us vivid personal characters, admirably as he draws conventional types. His truth to nature is positive; he has the eye of a Thoreau, and the pastoral fidelity which befits one who is not only the pupil of Milton and Keats, but of Theocritus and Wordsworth. He can treat broadly, and imaginatively withal, "the league-long roller thundering on the reef" and "the long wash of Australasian seas;" but his frequent over-elaboration led the way to a main fault of the younger schools.

While a poet cannot be too accurate, his method, to be natural, must seem unconscious. <small>Naturalness.</small> The virtue of a truth is spoiled by showing it off. Tennyson, the idylist, pauses at critical moments, not perhaps to moralize on the situation, but to make a picture suggesting the feeling which the action itself ought to convey. This practice, for a time so fascinating, has been carried to extremes. Now, in a class of his poems of which "Dora" is a fine example, he has shown that nothing can be more effective than a story simply told. A direct statement, through its truth, often has ex- <small>Force of a direct and simple method.</small> ceeding beauty,— the beauty, pathetic or otherwise, of perfect naturalness. You find it everywhere in the Scriptures; for example:—

> "I shall go to him, but he shall not return to me;"

and everywhere in Homer: —

> "A thousand fires burned in the plain, and by the side of each sate fifty in the gleam of blazing fire."

> "A deep sleep fell upon his eyelids, a sound sleep, very sweet, and most akin to death."

All genuine epics and ballads are charged with it, as in "The Children in the Wood:" —

> "No burial this pretty pair
> Of any man receives,
> Till Robin-redbreast piously
> Did cover them with leaves."

In the heroic vein, Arnold's "Sohrab and Rustum" has a primitive directness: —

> "So said he, and his voice released the heart
> Of Rustum; and his tears broke forth; he cast
> His arms around his son's neck, and wept aloud,
> And kissed him. And awe fell on both the hosts
> When they saw Rustum's grief."

The finest touch in Lady Barnard's ballad is the simplest, — that of the line,

> "For auld Robin Gray is kind unto me."

But I need not multiply such examples of the beauty of direct statement of unsophisticated truth. It is too rare a grace among the analytic and decorative poets.

When we come to the reflective poetry of nature, the broad effects of Wordsworth and Bryant are both true and imaginative, and therefore excellent realism. For Nature does not differentiate her beauties; she combines them. It is hard to better the truth "by her own sweet and cunning hand put on." Bryant's successors — Whittier, Lowell, Whitman, Lanier, Taylor — have great fidelity to nature. How can they help it, brought up in her own realm? Their touches are spontaneous, and that is everything. A city-bred poet is apt to strike false notes as soon as he hints at an intimacy with nature, and a false note is as quickly detected in poetry as in music, even by those who cannot sound the true one. As for truth to life — that depends on the poet's sympathetic perception. It was native to Burns; it was impossible with the self-absorbed Byron. Most poets, whether cockney or rustic, can draw only the types under their direct observation. Whitman's out-of-door poetry should be familiar to you. His admirers, including very authoritative judges at home and abroad, make almost every claim for him except that to which, in my opinion, he is entitled above other American poets. I know no other who surpasses him as a word-painter of nature. His eye is keen, his touch is accurate. No one depicts the American sky, ocean, forest, prairie, more characteristically or with a freer sense of atmosphere; no one is so inclusive of every object, living or inanimate, in the zones covered by our native land. His defects lie in his

Truth to visible Nature.

Excellence of the American school.

Whitman and Lanier.

theory of unvarying realism. Nature's poet must adopt her own method; and she hides the processes that are unpleasant to see or consider. Whitman often dwells upon the under side of things,— the decay, the ferment, the germination, which nature conducts in secret, though out of them she produces new life and beauty. Lanier, with equal fidelity, avoids — a refined and spiritual genius needs must avoid — this irritating mistake. His taste made him an open critic of the robust poet of democracy: but it is manifest that the two (as near and as different as Valentine and Orson) were moving in the same direction; that is, for an escape from conventional trammels to something free, from hackneyed time-beats to an assimilation of nature's larger rhythm,— to limitless harmonies suggested by the voices of her winds and the diapason of her ocean billows. The later portions of Whitman's life-work, his symphonies of "starry night," of death and immortality, have chords that would have thrilled Lanier profoundly.

In certain poems which have been humorously compared to "catalogues," Whitman supplies an example of the uselessness of a display of mere facts. Facts, despite Carlyle's eulogy upon them, are not "the one" and only "pabulum." They are the stones heaped about the mouth of the well in whose depth truth reflects the sky.[1]

True realism is not a statement of facts,

[1] "There is a way of killing truth by truths. Under the pretence that we want to study it more in detail, we pulverize the statue." — AMIEL.

I recall the words of Sir William Davenant, who wrote the feeblest of epics on a theory, yet preluded it with a chapter of noble prose wherein, among other fine discriminations, he says: "Truth, narrative and past, is the idol of historians (who worship a dead thing), and truth, operative and by its effects continually alive, is the mistress of poets, who hath not her existence in matter but in reason." A masterwork appeals, in time if not immediately, to the people at large as well as to the elect few, — to the former, doubtless, by its obvious intent and fidelity; to the critical, by its ideal and artistic truth; yet I think that the more esoteric quality is felt, if not comprehended, even by the masses, — that this makes, however vaguely and mysteriously, an impression upon their natures. Realism, in the sense of naturalism, is the firm ground of all the arts, but the poet, then, is not a realist merely as concerns the things that are seen. He draws these as they are, but as they are or may be at their best. This lifts them out of the common, or, rather, it is thus we get at the "power and mystery of common things." His most audacious imaginings are within the felt possibilities of nature. But the use of poetry is to make us believe also in the impossible. Raphael said that he painted "that which ought to be." And Browning writes: [nor a servile imitation.]

> "In the hall, six steps from us,
> One sees the twenty pictures — there's a life
> Better than life — and yet no life at all."

Lord Tennyson is reported as saying, with respect to certain contemporary writers: "Truth, as they understand it, is not the essential thing in poetry. For me verses have no other aim than to call to life nobler and better sentiments than we feel, and express in every-day life. If they can suggest pictures worthy of an artist's eye, so much the better." Even the first English writer upon the topic—George Puttenham, whose "Arte of English Poesie" was published anonymously in the year 1589—said that "Arte is not only an aide and coadjutor to nature in all her actions, but an alterer of them, so as by meanes of it her owne effects shall appear more beautiful or straunge and miraculous." And so there is nothing more lifeless, because nothing is more devoid of feeling and suggested movement, than servilely accurate imitation of nature. Moreover, in poetry as in all other art, a certain deviation from fact is not only justifiable, but required. Some things must be told or painted not as they are, but as they affect the eye or the imagination. The photograph reveals, indeed, the absolute position of the horse's legs at a given instant; by its aid the spokes of the revolving wheel are defined. Without doubt, art has learned most important facts through the photographic demonstration of actual processes; our animal- and figure-painters, our sculptors, can never repeat the absurd untruths which have become almost academic in the past. They will not,

It is vital with suggestion and interpretation.

Truth of relation to the human faculties.

and need not, however, go to the other extreme. To the human eye, with its halting susceptibilities, the horse and the wheel do not appear exactly as when caught by Mr. Muybridge's camera, and the artist's office is to present them as they seem to us. In the prosaic photograph they are struck with death: the idea of life, of motion, can only be conveyed by blending the spokes of the wheel as they are blended to the human vision, and by giving a certain unreality of grace to the speeding animal. Otherwise, you have the fact, which is not art.

Thus every workman must be a realist in knowledge, an idealist for interpretation, and the antagonism between realists and romancers is a forced one; and when any one rules the poet out of debate, as of course a feigner, he is in error, for the same law applies to all the arts. The true inquiry concerns the quality of the writer, his power of expression, the limits of his character. For no small and limited nature can enter into great passions and experiences. *Poetic truth is both realistic and ideal. — See, also, p. 145.*

It is a fine thing for a poet to express the life, feeling, ideal, of his own people; by so doing, he betters his chance of commending himself to after times. This is what the Greeks did, but in our century we find poet after poet exercising his skill upon reproductions, working the Grecian myths and legends over and over again in pseudo-classical lyrics, idyls, and dramas. The *Truth of environment.*

appeal of the loveliest and most successful *nova antica* — of a poem like "The Hamadryad" or "Œnone" — is to the æsthetic sense chiefly, and therefore in some measure restricted. After Lan-

<small>Raison d'être.</small> dor and Keats and Tennyson and Swinburne, our younger school cannot find a real need for this sort of thing. I remember my own chagrin, twenty years ago, when Mr. Lowell wrote a most judicious notice of one of my books, and failed to mention a blank-verse poem, with a classical theme, upon which I had expended the technical

<small>"Local flavor" not to be contemned.</small> skill and imagery at my command. On the other hand, he was more than kind to my native, if homely, American lyrics and ballads, written with less pains, yet more spontaneously; and he told me very frankly that he thought the simple home-fruit of more real significance than my attempt to reproduce some apple of the Hesperides. He was right, and I have not forgotten the lesson. With respect to another art, I wonder that the

<small>A home-field.</small> American sculptor does not still more frequently make a diversion from his imitations of the mediæval and the antique. What subjects he has close at hand, — such as a Greek, if he now could chance upon them, would handle with eagerness and truth! Surely our American workman, at labor and in repose, our young athletes, our beasts of the forest and of the field, are available models; and Ward's "Indian Hunter," Donoghue's "The Boxer," and Tilden's "The Ball-Thrower," at least

convey their suggestion of what should and will be done. There is a certain lack of sincer- <small>Sincerity.</small> ity, despite their artistic beauty, in the foreign and antique exploits of many poets and artists; and lack of sincerity is always lack of truth. But, while they should favor their own time, they must avoid expression of its transient passions and characteristics. Seize upon the essential, lasting traits, and let the others be accessory. If the general spirit of the time be not embodied, a work is soon out of date.

Against all this, the widest freedom is permitted to that chartered libertine,— the poet's <small>But nothing is forbidden to</small> imagination. Nature and the soul being <small>the imagination, and a</small> the same forever, we care nothing for <small>poet may follow his</small> Shakespeare's anachronisms and impossi- <small>bent.</small> ble geography; we find nothing strange and unnatural in his assembly of mediæval fays and antique heroes and amazons, of English clowns and mechanics in Grecian garb, all commingled to enact a fantastic marvel of comedy and poesy in the palace and forests of a "Midsummer Night's Dream." We confess the poet's witchcraft, and ourselves are of the blithe company,— denizens of an enchanted land, where everything has the truth of possibility. A conception is not vitiated by the most novel form it may assume, provided that this be artistic and not artificial. For art, as Goethe and Haydon have said, is art because it is not nature. That method is most true which, invoking the force of nature, directs it by its own device; just as, in mechanics, the screw-propeller is more than the equivalent of the fish's flukes or the bird's wing. Our delight in art proceeds from a knowledge that it is not inevitable, but designed; a human, not a natural, creation; the truth of nature's capabilities, seen by man's imagination, captured by the human hand, expressed and illumined when our Creator, intrusting his own wand to us, bids us test its power ourselves.

<small>Art has a truth of its own.</small>

What is called descriptive poetry never can be very satisfying, since the painter is so much more capable than the poet of transferring the visible effects of nature, — those addressed to the eye. I suppose it is out of the power of one not reared in England, and in that very part of England which lies between Derwentwater and the Wye, to comprehend thoroughly the truth and beauty of Wordsworth's pastoral note and landscape. Neither can a foreigner rightly estimate the American idylists; the New World scenery and atmosphere are so different from the European that they must be seen before their quality can be felt. Aside from this limitation, the poet expresses what he finds in nature, to wit, that which answers to his own needs and temper. Her interpretation has been, it may almost be said, a special function of the century now closing. Nature moved Coleridge to eloquence, rhapsody,

<small>The poet inferior to the painter in depicting nature;</small>

<small>but unsurpassed as her subjective interpreter.</small>

worship, and, as an artist, to imaginative mysticism. Heine, Longfellow, Swinburne, have read the secret of the sea. To Landor, Emerson, and Lowell the tree is animate; in their presence the flower has rights: they would not fell the one nor pluck the other. But there were two English poets whose respective temperaments answered perfectly to the two conditions of nature embraced in Lord Bacon's profound observation, that "In nature things move violently *to* their place and calmly *in* their place." Byron's fitful genius was stirred by her violence of change. The rolling surges, the tempest, the live thunder leaping from peak to peak, mated the restlessness of a spirit charged with their own intensity of motion and desire. Wordsworth felt the sublimity of the repose that lies on every height, of nature's ultimate subjection to law. His imagination comprehended her reserved forces; and before his time her deepest voice had no apt interpreter, for none had listened with an ear so patient as his for mastery of her language. His announcement that

<blockquote>
" he who feels contempt
For any living thing, hath faculties
Which he has never used,"
</blockquote>

was like a revelation. That he had purged himself of all such baseness was his absolute conviction; in such matters he was a kind of Gladstone among the poets of his day. Therefore, self-contemplation, or, to be more exact, the transcription of nature's effect upon himself, seemed to him a sane, even a sacred vocation. In fact, a lofty, if not inventive, imagination, and

> "An eye made quiet by the power of harmony,"

gave him for this faith a warrant which all his ponderous homiletics could not render null. As he let *The modern return to nature.* "the misty mountain winds" blow on him, he was nature's living oracle. And the world soon yielded to the force of that "pathetic fallacy" which has imparted to modern thought a distemper and a compensation: the refuge, be it real or illusionary, still left to us, and so compulsive that neither reason nor science can quite rid us of it when face to face with nature, — when soothed by the sweet influences of our mother Earth. It is true, in Landor's words, that

> "We are what suns and winds and waters make us;
> The mountains are our sponsors, and the rills
> Fashion and win their nursling with their smiles."

But Ruskin avers that the illusion under which we fondly believe nature to be the sympathetic participator of our sentiment or passion, and which he terms the pathetic fallacy, is incompatible with a clear-seeing acceptance of the truth of things.

Now, that there is a solace — a companionship — *The "pathetic fallacy."* found in nature none can doubt. It is as old as the fable of Antæus. Primitive races feel it so strongly that they inform all natural objects with sentient individual lives; our more advanced intelligence conceives of a universal spirit that comprehends and soothes Earth's children. In our own youth, nature haunts us "like a passion;" and as concerning the youth of a race we "cannot paint what then" we were, in mature years each of us can say, —

> "And I have felt
> A presence that disturbs me with the joy
> Of elevated thoughts; a sense sublime
> Of something far more deeply interfused,
> Whose dwelling is the light of setting suns,
> And the round ocean and the living air,
> And the blue sky, and in the mind of man.
> A motion and a spirit, that impels
> All thinking things, all objects of all thought,
> And rolls through all things."

This has never been expressed so well as in Wordsworth's elevated phrases. They must always be cited. But a disenchantment is at last upon us, and we are sternly questioning our reason. Is not nature's apparent sympathy, we ask, a purely subjective illusion? The old belief, the new doubt, are well conveyed in the early and later treatment of a favorite theme, — the moaning of a sea-shell held to the ear. In Landor's "Gebir" we have it thus: — *(Expressions of the old feeling and the new doubt.)*

> "But I have sinuous shells of pearly hue;
> . . . .
> Shake one and it awakens, then apply
> Its polished lips to your attentive ear,
> And it remembers its august abodes,
> And murmurs as the ocean murmurs there."

(The shell's murmur, as idealized by Landor and Wordsworth.)

Landor complained that Wordsworth stole his shell, and "pounded and flattened it in his marsh" of "The Excursion":—

> "I have seen
> A curious child, who dwelt upon a tract
> Of inland ground, applying to his ear
> The convolutions of a smooth-lipped shell;
> To which, in silence hushed, his very soul
> Listened intensely; and his countenance soon
> Brightened with joy; for from within were heard
> Murmurings, whereby the monitor expressed
> Mysterious union with its native sea."

Byron acknowledged his obligations to "Gebir" for his lines in "The Island," beginning,—

> "The Ocean scarce spake louder with his swell,
> Than breathes his mimic murmurer in the shell."

And now, as we near the close of the century which "Gebir" initiated, Eugene Lee-Hamilton devotes one of his remarkable sonnets to this same murmur of the shell, and I cannot find a more poetic, more impassioned recognition of the veil which modern doubt is drawing between our saddened eyes and the beautiful pathetic fallacy:—

And as now reinterpreted by Lee-Hamilton.

> "The hollow sea-shell which for years hath stood
> On dusty shelves, when held against the ear
> Proclaims its stormy parent; and we hear
> The faint far murmur of the breaking flood.
> We hear the sea. The sea? It is the blood
> In our own veins, impetuous and near,
> And pulses keeping pace with hope and fear
> And with our feelings' ever-shifting mood.

> " Lo! in my heart I hear, as in a shell,
>   The murmur of a world beyond the grave,
>   Distinct, distinct, though faint and far it be.
> Thou fool! this echo is a cheat as well, —
>   The hum of earthly instincts; and we crave
>   A world unreal as the shell-heard sea."

How beautiful this ecstasy of disenchantment, — beautiful in its sad sincerity, — and yet how piteous! Here is a fine spirit, for the moment baffled, heroically demanding the truth, the truth. More trustfully leaving the future to "the Power that makes for good," Lowell also confronts the scientific analysis of our attitude toward nature : — *Truth before all,*

> " What we call Nature, all outside ourselves,
>   Is but our own conceit of what we see,
>   Our own reaction upon what we feel;
>   The world's a woman to our shifting mood,
>   Feeling with us, or making due pretence;
>   And therefore we the more persuade ourselves
>   To make all things our thoughts' confederates,
>   Conniving with us in whate'er we dream."

The poet, to be aware of this, must have drifted quite away from the antique point of view. The Greek certainly made nature populous with dryads, oreads, naiads, and all the daughters of Nereus; but these had a joy and, like Jaques, a melancholy of their own, not those of common mortals. Doubtless the Greek felt the charm of the hour when twilight descended on his valley, but not the pensive suggestions of the Whence and Whither which it excites in you and me. "No young man," *though with it disenchantment.*

said Hazlitt, "ever thinks he shall die." He recognizes death, but it concerns him not. The Greek accepted it as a natural process; he yielded to nature; we adjure her, as Manfred adjured his spirits, and fain would compel her to our service and demand her to surrender the eternal secret.

Nature, even in her most tranquil mood, is palpitant with motion, in view of which Humboldt was at times a poet. Motion is life, and therefore fellowship. Herein lies the spell of the sea, which has mastered Heine and Shelley and every poetic soul. Its perpetual change, eternal endurance — these image both life and immortality; its far-away vessels moving to unknown climes, its unbounded horizon suggesting infinity, buoy the imagination, and thence come human passion and thoughts "too deep for tears." We have conquered it, and it is the modern poet's comrade, as it was the ancient's fear and marvel. But what is the sea? Tennyson's "still salt pool, lock'd in with bars of sand," would be an ocean to a man reduced to insect size,— a stretch of water, infused with salt, and roughened into wavelets by the air that moves across it. We have learned that the effect of the sea, of a prairie, of a mountain, is purely relative. One of the latest "Atlantic" novelists, with youth's contemporaneousness, realizes both the fact and the dream. Her lovers are watching "a big, red, distorted moon above the illimitable palpitating waste" of the ocean: —

[marginal note: Why Nature yields solace and companionship.]

"A waning moon is so melancholy," said Felicia, looking at it with wide, soft eyes that had grown melancholy, too. "I wonder why?"

"I don't see that it is melancholy," Grafton declared.

"No, I suppose not," she rejoined. "I dare say you see a planet which suggests to you apogee, or perigee, or something wise. I see only the rising moon, and it seems to me particularly ominous to-night. I am afraid. Something unexpected — perhaps something terrible — is going to happen."

You will note, by the way, that our débutante is scientifically accurate upon a matter in respect to which many a good writer has gone wrong. She sees the moon where it should be of an evening in its third quarter,— to wit, rising in the east. Giving the author of "Felicia" credit for this unusual feat, I believe that reason never can greatly lessen the influence of nature upon our feelings, and this in spite of her stolid indifference, her want of compassion, her stern laws, her unfairness, unreason, and general unmorality. To the last, man will be awed by the ocean and saddened by the waning moon, and will find the sun-kissed waves sparkling with his joy, and the stars of even looking down upon his love. One may conceive, moreover, that before a vast and various landscape we are affected by the very presence of divinity revealed only in his works; that, face to face with such an expanse of nature, we recognize more of a pervading spirit than when more closely pent:

and this in spite of our scientific readjustment.

as in a house of worship, with a host of others like ourselves, we have more of him incarnate in humanity; whence comes a strange exaltation, and at times almost a yearning to be reabsorbed in the infinite being from which our individual life has sprung.

<small>Nature the sovereign of modern art and song.</small>
The aspect and sentiment of nature, more than other incentives to mental elevation, have supplied a motive to the artistic expression of the last half century. In the domains of the painter and the poet, and on both sides of the Atlantic, the idealization of nature has been, as never before, supreme. Never has she been portrayed on canvas as by Turner and his successors; never has she received such homage in song as that of the English and American poets from the time of Wordsworth. Two significant advantages confirmed Wordsworth's influence: first, that of longevity, which, in spite of the ancient proverb, is the best gift of the gods to an originative leader; second, the fact that, with brief exceptions, he made verse his only form of expression. No wonder that he produced an "ampler body" of good poetry — and of prosaic verse as well — than "Burns, or Keats, or Manzoni, or Heine." But in this country, also, the force of nature has been sovereign, since Bryant first gave voice to the spirit of the glorious forest and waters of a relatively primeval land. During an idyllic yet speculative period, the maxim that "the proper study of mankind is man" has for many reasons been almost in abeyance. At last it is again

A LIFE-SCHOOL DEMANDED 211

evident that we cannot live by bread alone, even at the hands of the great mother. There is a longing and a need for emotion excited by action and life, for a more impassioned and dramatic mode,—that of a figure-school, so to speak, in both poesy and art. Not to "fresh woods and pastures new," but to human life with its throes and passions and activity, must the coming poet look for the inspirations that will establish his name and fame.

<small>Her triumph too prolonged. Cp. "Poets of America": pp. 464-466.</small>

In my censure of didacticism I used that word in the usually adopted sense. Its radical meaning is not to be dismissed so lightly. If there is a base didacticism false to beauty and essentially commonplace, there is a nobly philosophic strain which I may call the poetry of wisdom. There is an imagination of the intellect, and its utterance is of a very high order, — often the prophecy of inspiration itself.

<small>Philosophic truth. The higher didacticism.</small>

Were this not so, we should have to reverse time's judgment of intellectually poetic masterpieces from which have been derived the wisdom and the rubrics of many lands. Shall we rule out the lofty voice of the Preacher, whose lesson that all save the fear of God is vanity has been reaffirmed by a cloud of witnesses, down to the chief of imaginative homilists in our own time? Whether prose or verse, I know nothing grander than Ecclesiastes in its impassioned survey of mortal pain and

<small>Ecclesiastes.</small>

pleasure, its estimate of failure and success; none of more noble sadness; no poem working more indomitably for spiritual illumination. Shall we rule out the elegies of Theognis or the mystic speculations of Empedocles, celebrant of the golden age and declarer of the unapproachable God? And who would lay rude hands upon the poet who concerned himself with the universe, surpassing all other Latins in intellectual passion and dignity of theme? The rugged "De Rerum Natura" of Lucretius seems to me as much greater than the Æneids as fate and nature are greater than the world known in that day. Whether his science was false or true, — and meanwhile you know that the atomic theory is once more in vogue, — he essayed "no middle flight," but soared upon the philosophy of Epicurus to proclaim the very nature of things; meditating which, as he declared, the terrors of the mind were dispelled, the walls of the world parted asunder, and he saw things "in operation throughout the whole void." What shall we do with Omar Khayyám, at least with that unique paraphrase of his "Rubáiyát" which has impressed the rarest spirits of our day, and has so inspired the wondrous pencil of Elihu Vedder, our American Blake? And what of "In Memoriam"? The flower of Tennyson's prime is distinctly also the representative Victorian poem. It transmits the most characteristic religious thought of our intellectual leaders at

[margin: The Grecian sages.]
[margin: Lucretius.]
[margin: Omar.]
[margin: The wise imagination of our recent time.]

the date of its production. We have no modern work more profound in feeling, more chaste in beauty, and none so rich with the imaginative philosophy of the higher didacticism. Browning's precepts, ratiocination, morals, are usually the weightier matters of his law. Take from Emerson and Lowell their sage distinctions, their woof of shrewdest wisdom, and you find these so closely interwoven with their warp of beauty that the cloth of gold will be ruined. Like Pope and Tennyson, they have the gift of "saying things," and in such wise that they add to the precious currency of English discourse.

The mention of Pope reminds me that he is the traditional exemplar of the didactic heresy, so much so that the question is still mooted whether he was a poet at all. *Pope, as the chief of English moralist-poets.* As to this, one can give only his own impression, and my adverse view has somewhat changed, — possibly because we grow more sententious with advancing years. Considering the man with his time, I think Pope was a poet: *The question concerning his inherent gift.* one whose wit and reason exceeded his lyrical feeling, but still a poet of no mean degree. Assuredly he was a force in his century, and one not even then wholly spent. His didacticism was inherent in the stiff, vicious, Gallic drum-beat of his artificial style — so falsely called "classical," so opposed to the true and live method of the antique — rather than in his genius and quality. It is impossible that one with so marked a poetic temperament, and using verse withal as almost his sole mode of expression, should not have been a poet. In the manner of his time, how far above his rivals! Every active literary period has one poet at least. To me he seems like the tree which, pressed hard about by rocks, adorns them and struggles into growth and leafage. A fashion of speech mastered him, but he refined it and made it effective, the wonder being that he did so much with it. All admit that Cowper was a poet and the pioneer of a noble school. But he was as didactic as Pope; his vantage lay in a return to natural diction and flexible rhythm. A free vehicle of expression sets free the imagination. Again, there are forms still in use, and natural, as we say, to the genius of our language, in which Pope's resources were sufficient for the display of lasting thought and emotion. "The Universal Prayer" and "The Dying Christian to his Soul" equal the best of Cowper's lyrics. "The Rape of the Lock," still the masterwork of patrician verse, shows what its author could do with a subject to which his grace, wit, and spirit were exactly suited. The passionate intensity of "Eloisa to Abelard" lifts that epistle far above the wonted liberties of its formal verse. Looking at the man, Pope, that fiery, heroic little figure, that vital, electric spirit pitiably encaged, — defying and conquering his foes, loving, hating, questioning, worshipping, — I see the poet. However, if you care to realize how much more difference there is in the methods than in the contemplative gifts of certain bards, amuse yourselves by translating Pope, Tennyson, Emerson, Browning, into one another's measures and styles, and you will find the result suggestive. Autre temps, autres mœurs.

Three, at least, of these poets have at times a delicious humor and fancy, as in "The Rape of the Lock," "The Talking Oak," "Will Waterproof's Lyrical Monologue," "The Pied Piper," etc. Humor, in the sense of fun, is doubtless another lyrical heresy. But humor is the overflow of genius, — the humor compounded of mirth and pathos, of smiles and tears, — and in the poems cited, and in Thackeray's ballads, it speaks for the universality of the poet's range. While certain notes in excess are fatal to song, in due subordination they supply a needful relief, and act as a fillip to the zest of the listener. Humor as a poetic element.

In speaking, as I have, of measures and diction suited to the English language, it must be with reservation. That language has advanced of late so rapidly from the simple to the complex, that it seems ready to assimilate whatever is most of worth in the vocabularies and forms of many tongues. In Pope's time it had thrown away, "like the base Indian," half the riches bequeathed by Chaucer and the dramatists; nevertheless, an age of asceticism often leads to one of prodigal vigor. It required long years after Pope, and a French Revolution, to renew the affluence of English letters, but if the process was slow Eclectic genius of our English tongue.

it was effectual. Another century has passed; our language, in turn, is giving increment to the Continental tongues, and the need of an artificial Volapük may soon disappear before this eclectic universality.

The highest wisdom — that of ethics — seems *Truth of ethical insight.* closely affiliated with poetic truth. A prosaic moral is injurious to virtue, by making it repulsive. The moment goodness becomes tedious and unideal in a work of art, it is not real goodness; the would-be artist, though a very saint, has mistaken his form of expression. On the other hand, extreme beauty and power in a poem or picture always carry a moral: they are inseparable from a certain ethical standard; while vice suggests a depravity. Affected conviction, affection of any kind, and even sincere conviction inartistically set forth, are vices in themselves, — are antagonistic to truth. But the cleverest work, if openly vicious, has no lasting force. A meretricious play, after the first rush of the baser sort, is soon performed to empty boxes. Managers know this to be so, and what is the secret of it? Simply, that to cater to a sensual taste incessant novelty is required. *Why baseness nullifies the force of art.* Vice admits of no repose; its votary goes restlessly from one pleasure to another. Thus no form of vicious art bears much repetition: it satiates without satisfying; besides, any one who cares for art at all has some sort of a moral standard. He violates it himself, but does not care to see it violated in art as if upon principle.

An obtrusive moral in poetic form is a fraud on
its face, and outlawed of art. But that all *Enduring poetry always makes for good.*
great poetry is essentially ethical is plain
from any consideration of Homer, Dante,
and the best dramatists and lyrists, old and new.
Even Omar, in proud recognition of the immutability of the higher powers, chants a song without fear
if without hope. The pagan Lucretius, confronting
sublimity, found no cause to fear either the gods or
the death that waits for all things. A glimpse of
the knowledge which is divine, an approach to the
infinite which makes us confess that "an undevout
astronomer is mad," inspire the "De Rerum Natura."
The poet sat in the darkness before dawn. He
would report no vision which he did not see. Like
Fitzgerald's Omar he seems to confess, with the
epicureanism that after all is but inverted stoicism,
and with unfaltering truth,—

> "Up from Earth's Centre through the Seventh Gate
> I rose, and on the Throne of Saturn sate,
>   And many a Knot unravell'd by the Road;
> But not the Master-knot of Human Fate."

Poetry, in short, as an ethical force, may be either
iconoclastic or constructive, nor dare I *A noble scepticism.*
say that the latter attribute is the greater,
for the site must be cleared before a new edifice can
be raised. Herein consists the moral integrity of
Lucretius and Omar. They rebelled against the
superstitions of their periods. Better a self-respecting confession of ignorance, a waiting for some voice

from out the void, than a bowing down to stone images or reverence for a false prophet. Critics are still to be found who look upon a modern poet — in his lifetime almost an outlaw — as a splendid lyrical genius gone far astray. Of course I refer to Shelley.

Percy Bysshe Shelley. The world is slowly learning that Shelley's office, if any need be ascribed to him save that of charming the afterworld with song, was ethical. As an iconoclast, he rebelled against tyranny and dogma. His mistakes were those of poetic youth and temperament, and he grew in love, justice, pity, according to his light. He groped in search of some basis for construction, but died in *The false standards of criticism applied to his life and works.* what was still his formative period. Yet we see sage and elderly moralists applying to Shelley the tests of their own mature years and modern enlightenment, and holding a sensitive and passionate youth to account as if he were an aged philosopher.[1] Even Matthew Arnold, despite his fine recognition of that transcendent lyrist, did not quite avoid this attitude. Professor Shairp assumed it altogether. With respect to the poetry of nature, I can refer you to no more sugges-

---

[1] Some reviewer, alluding to the discussion of Hawthorne's career, has said with much intelligence that the romancer was first of all, by choice and genius, an artist, and that his politics, ethics, etc., are matters quite subordinate in any estimate of him. It is well, then, to aver that Shelley was, before all else and marvellously, a *poet*, and that the rapid experiences of his young life — which ended, indeed, before the age of mature convictions — are of importance merely as they affected what we have inherited of his beautiful lyric and dramatic creations.

tive critic, for he was a Wordsworthian, and all his discourse leads up to Wordsworth as the greatest, because the most contemplative, of nineteenth-century poets. Otherwise he was an extreme type of the class which Arnold had in mind when he said, "We must be on our guard against the Wordsworthians, if we want to secure for Wordsworth his due rank as a poet." His utter failure to see the force of a blind revolt like Shelley's, in the evolution of an ultimately high morality, was inexcusable. A more striking example of faulty criticism could hardly be given. Shelley is not to be measured by his conduct of life nor by his experimental theories, but rather, as Browning estimates him, with every allowance for his conditions and by his highest faculty and attainment.

But the most thoughtful and extended of rhythmical productions in the purely didactic method is of less worth, taken as poetry, than any lyrical trifle — an English song or Irish lilt, it may be — that is spontane- *Poetic truth, above all, is hostile to the commonplace and unimaginative,* ous and has quality. The disguises of the commonplace are endless; we are always meeting the old foe with a new face. A fashionable diction, tact, taste, the thought and manner of the season, set them off bravely; but they soon will be flown with the birds of last year's nests. Of such are not the works whose wisdom is imaginative, whether the result of intuition or reflection, or of both combined. These "large utterances" of intellectual and moral truth show that nothing is impossible, no domain is forbidden, to the poet, that no thought or fact is incapable of ideal treatment. The bard may proudly forego the office of the lecturer, such as that exercised in this discourse, which is by intention didactic and plainly inferior to any fine example of the art to which its comment is devoted. Yet the new learning doubtless will inspire more of our expression in the near future, since never was man so apt in translation of nature's oracles, and so royally vouchsafed the freedom of her laboratory, as in this age of physical investigation. Accepting the omen, we make, I say, another claim for the absolute liberty of art. Like *Gaspar Becerra*, the artist must work out his vision in the fabric nearest at hand. His theme, his method, shall be his own: always with the passion for beauty, always with an instinct for right. No effort to change the natural bent of genius was ever quite successful, though such an effort often has spoiled a poet altogether.

<small>but alert in each new wonderland.</small>

This brave freedom alone can breed in a poet the catholicity which justifies Keats' phrase, and insures for his work the fit coherence of beauty and truth. The lover of beauty, in Emerson's "Each and All," marvels at the delicate shells upon the shore:—

<small>The poet's final recognition of beauteous verity.</small>

> "The bubbles of the latest wave
> Fresh pearls to their enamel gave;

> ". . . . .
> "I wiped away the weeds and foam,
> I fetched my sea-born treasures home;
> But the poor, unsightly, noisome things
> Had left their beauty on the shore,
> With the sun, and the sand, and the wild uproar."

Disappointed, he forswears the pursuit of beauty, and declares:—

> "I covet truth;
> Beauty is unripe childhood's cheat;
> I leave it behind with the games of youth."

But, even as he speaks, the ground-pine curls its pretty wreath beneath his feet, "running over the club-moss burrs;" he scents the violet's breath, and therewithal

> "Over me soared the eternal sky,
> Full of light and deity;
> . . . .
> Beauty through my senses stole;
> I yielded myself to the perfect whole."

This recognition, at which the idealist arrives, of the intertransmutations of beauty and truth, is a kind of natural piety, and renders the labor of the poet or other "artist of the beautiful" a proper form of worship. His heart tells him that this is so: it is lightest when he has worked at his craft with diligence and accomplishment; it is light with a happiness which the religious say one can know only by experience. The piety of his labor is not yet sufficiently comprehended; even the poet, having listened all his life to other tests of sanctification, often mistrusts his own conscience, looks

Labor est etiam ipsa pietas.

upon himself as out of the fold, and is sure only that he must "gang his ain gait," however much he suffers for it in this world or some other.

Thus a dividing line has been drawn from time immemorial betwixt the conventional and the natural worshippers, betwixt the stately kingdom of Philistia and the wilding vales and copses of that Arcadia which some geographers have named Bohemia. The mistake of the Arcadian is that he virtually accepts a standard not of his own establishment; he is impressed by a traditional conception of his Maker, regards it as fixed, will have none of it, and sheers off defiantly. If rich and his own master, he becomes a pagan virtuoso. If one of the struggling children of art and toil, then, —

Arcadian non-conformity.

> " Loving Beauty, and by chance
> Too poor to make her all in all,
> He spurns her half-way maintenance,
> And lets things mingle as they fall."

This is the way in Arcadia, and it has its pains and charm, — as I well know, having journeyed many seasons in that happy-go-lucky land of sun and shower, and still holding a key to one of its entrance-gates. Its citizenship is not to be shaken off, even though one becomes naturalized elsewhere.

Now the artist not only has a right, but it is his duty, to indulge an anthropomorphism of his own. In his conception the divine power must be the supreme poet, the matchless artist, not only the transcendency, but the

The God of truth is no less the God of beauty, joy, and song.

immanence of all that is adorable in thought, feeling, and appearance. Grant that the Creator is the founder of rites and institutes and dignities; yet for the idealist he conceived the sunrise and moonrise, the sounds that ravish, the outlines that enchant and sway. He sets the colors upon the easel, the harp and viol are his invention, he is the model and the clay, his voice is in the story and the song. The love and the beauty of woman, the comradeship of man, the joy of student-life, the mimic life of the drama as much as the tragedy and comedy of the living world, have their sources in his nature; nor only gravity and knowledge, but also irony and wit and mirth. Arcady is a garden of his devising. As far as the poet, the artist, is creative, he becomes a sharer of the divine imagination and power, and even of the divine responsibility.

VII.

IMAGINATION.

IT is worth while to reflect for a moment upon the characteristics of recent poetry. Take, for example, the verse of our language produced during the laureateship of Tennyson, and since the rise, let us say, of Longfellow and his American compeers.

Qualities of modern verse.

In much of this composition you detect an artistic convergence of form, sound, and color; a nice adjustment of parts, a sense of craftsmanship, quite unusual in the impetuous Georgian revival,— certainly not displayed by any poets of that time except those among whom Keats was the paragon and Leigh Hunt the propagandist. You find a vocabulary far more elaborate than that from which Keats wrought his simple and perfected beauty. The conscious refinement of our minor lyrists is in strong contrast with the primitive method of their romantic predecessors. Some of our verse, from "Woodnotes" and "In Memoriam" and "Ferishtah's Fancies" down, is charged with wholesome and often subtile thought. There has been a marked idyllic picturesqueness, besides a variety of classical and Preraphaelite experiments,

Its conscious refinement and vocabulary.

and a good deal of genuine and tender feeling. Our leaders have been noted for taste or thought or conviction,—often for these traits combined. But we obtain our average impression of a literary era from the temper of its writers at large. Of late our clever artists in verse—for such they are—seem with a few exceptions indifferent to thought and feeling, and avoid taking their office seriously. A vogue of light and troubadour verse-making has come, and now is going as it came. Every possible mode of artisanship has been tried in turn. The like conditions prevail upon the Continent, at least as far as France is concerned; in fact, the caprices of our minor minstrelsy have been largely the outcome of a new literary Gallomania.

<small>Cp. "Victorian Poets": p. 477; and "Poets of America": pp. 458-460.</small>

Now, I think you will feel that there is something unsatisfactory, something much less satisfactory than what we find in the little prose masterpieces of the new American school; that from the mass of all this rhythmical work the higher standard of poetry could scarcely be derived. To be sure, it is the providential wont of youth to be impressed by the latest models, to catch the note of its own morntime. Many know the later favorites by heart, yet perhaps have never read an English classic. We hear them say, "Who reads Milton now, or Byron, or Coleridge?" It is just as well. Otherwise a new voice might not be welcomed,—would have less chance to gain a hear-

<small>Something more is needed to confer distinction.</small>

ing. Yet I think that even the younger generation will agree with me that there are lacking qualities to give distinction to poetry as the most impressive literature of our time; qualities for want of which it is not now the chief force, but is compelled to yield its eminence to other forms of composition, especially to prose fiction, realistic or romantic, and to the literature of scientific research.

If you compare our recent poetry, grade for grade, with the Elizabethan or the Georgian, I think you will quickly realize that the characteristics which alone can confer the distinction of which I speak are those which we call Imagination and Passion. Poetry does not seem to me very great, very forceful, unless it is either imaginative or impassioned, or both; and in sooth, if it is the one, it is very apt to be the other.

"The two middle pillars upon which the house stood."

The younger lyrists and idylists, when finding little to evoke these qualities, have done their best without them. Credit is due to our craftsmen for what has been called "a finer art in our day." It is wiser, of course, to succeed within obvious limits than to flounder ambitiously outside them. But the note of spontaneity is lost. Moreover, extreme finish, adroitness, graces, do not inevitably betoken the glow of imaginative conception, the ecstasy of high resolve.

If anything great has been achieved without exercise of the imagination, I do not know of it. I am

referring to striking productions and achievements,
Anticipatio quædam deorum. not to acts of virtue. Nevertheless, at the last analysis, it might be found that imagination has impelled even the saints and martyrs of humanity.

Imagination is the creative origin of what is fine, not in art and song alone, but also in all forms of action, — in campaigns, civil triumphs, material conquest. I have mentioned its indispensability to the scientists. It takes, they surmise, four hundred and ninety years for the light of Rigel to visit us. Modern imagination goes in a second to the darkness beyond the utmost star, speculates whether the ether itself may not have a limiting surface, is prepared to see at any time a new universe come sailing from the outer void, or to discover a universe within our own under absolutely novel conditions. It posits molecules, atomic rings; it wreaks itself upon the ultimate secrets of existence. But in the practical world our men of action are equally, though often unwittingly, possessed by it. The imagination of inventors, organizers, merchant princes, railway kings, is conceptive and strenuous. It bridges rivers, tunnels mountains, makes an ocean-ferry, develops the forces of vapor and electricity, and carries each to swift utility; is already picturing an empery of the air, and doubtless sighs that its tangible franchise is restricted to one humble planet.

If the triumphs of the applied imagination have more and more engrossed public attention, it must be remembered that its exhibitors, accumulating wealth, promote the future structures of the artist and poet. In the Old World this has been accomplished through the instrumentality of central governments. In a democracy the individual imagination has the liberty, the duty, of free play and achievement. Therefore, we say that in this matter our republicanism is on trial; that, with a forecast more exultant, as it is with respect to our own future, than that of any people on earth, our theory is wrong unless through private impulse American foundations in art, learning, humanity, are not even more continuous and munificent than those resulting in other countries from governmental promotion. The executive Imagination.

As for the poetic imagination, as distinguished from that of the man of affairs, if it cannot parcel out the earth, it can enable us to "get along just as well without it," — and this by furnishing a substitute at will. There is no statement of its magic so apt as that of our master magician. It "bodies forth the forms of things unknown," and through the poet's pen Imagination of the poet.

> "Turns them to shapes, and gives to airy nothing
> A local habitation and a name."

I seldom refer to Shakespeare in these lectures, since we all instinctively resort to him as to nature itself; his text being not only the chief illustration of each phrase that may arise, but also, like nature, presenting all phases in combi- Shakespeare the preëminent exemplar.

nation. It displays more of clear and various beauty, more insight, surer descriptive touches, — above all, more human life, — than that of any other poet; yes, and more art, in spite of a certain constructive disdain, — the free and prodigal art that is like nature's own. Thus he seems to require our whole attention or none, and it is as well to illustrate a special quality by some poet more dependent upon it. Yet if there is one gift which sets Shakespeare at a distance even from those who approach him on one or another side, it is that of his imagination. As he is the chief of poets, we infer that the faculty in which he is supereminent must be the greatest of poetic endowments. Yes: in his wonderland, as elsewhere, imagination is king.

There is little doubt concerning the hold of Shakespeare upon future ages. I have sometimes debated whether, in the change of dramatic ideals and of methods in life and thought, he may not become outworn and alien. But the purely creative quality of his imagination renders it likely that its structures will endure. Prehistoric Hellas is far removed from our experience; yet Homer, by force of a less affluent imagination, is a universal poet to-day, — to-day, when there is scarcely a law of physics or of art familiar to us that was not unknown to Homer's world. Shakespeare's imagination is still more independent of discovery, place, or time. It is neither early nor late, antiquated nor modern; or, rather, it is always modern

"Not of an age, but for all time."

and abiding. The beings which he creates, if suddenly transferred to our conditions, would make themselves at home. His land is one wherein the types of all ages meet and are contemporary. He created beings, and took circumstances as he found them; that is, as his knowledge enabled him to conceive of them at the time. The garb and manners of his personages were also a secondary matter. Each successive generation makes the acquaintance of these creatures, and troubles itself little about their fashions and acquirements. Knowledge is progressive, communicable: the types of soul are constant, and are sufficient in themselves.

It does no harm, as I said at the outset of this course, for the most advanced audience to go back now and then to the primer of art, — to think upon the meaning of an elementary term. Nor is it an easy thing to formulate clear statements of qualities which we instantly recognize or miss in any human production, and for which we have a ready, a traditional, nomenclature. So, then, what is the artistic imagination, that of one who expresses his conceptions in form or language? I should call it a faculty of conceiving things according to their actualities or possibilities, — that is, as they are or may be; of conceiving them clearly; of seeing with the eyes closed, and hearing with the ears sealed, and vividly feeling, things which exist only through the will of the artist's genius. Not only of

<small>Definition of the artistic Imagination.</small>

conceiving these, but of holding one's conceptions so well in mind as to express them,— to copy them, — in actual language or form.

The strength of the imagination is proportioned, in fact, to its definiteness, and also to the stress of its continuance,— of the memory which prolongs its utilization. Every one has more or less of this ideal faculty. The naturalness of children enables us to judge of their respective allotments. A mother knows which of her brood is the imaginative one. She realizes that it has a rare endowment, yet one as perilous as "the fatal gift of beauty." Her pride, her solicitude, are equally centred in that child. Now the clearer and more self-retentive this faculty, the more decided the ability of one in whom it reaches the grade at which he may be a designer, an artist, or a poet.

How to gauge its strength.

Let us see. Most of us have a sense of music. Tunes of our own "beat time to nothing" in the head. We can retain the theme, or opening phrase, at least, of a new composition that pleases us. But the musician, the man of genius, is haunted with unbidden harmonies; besides, after hearing a difficult and prolonged piece, he holds it in memory, perhaps can repeat it,— as when a Von Bülow repeats offhand an entire composition by Liszt. Moreover, his mind definitely hears its own imaginings; otherwise the sonata, the opera, will be confused and inferior. Again: most of us, especially when nervous or half asleep, find the "eyes make

Clearness and retention.

pictures when they are shut." Faces come and go, or change with startling vividness. The face that comes to a born painter does not instantly go; that of an angel is not capriciously transformed to something imp-like. He sees it in such wise that he retains it and can put it on his canvas. He has the clear-seeing, the sure-holding, gift which alone is creative. It is the same with the landscape-painter, the sculptor, the architect. Artistic ability is coördinate with the clearness and staying-power of the imagination.

More than one painter has declared that when a sitter was no longer before him, he could still lift his eyes, and see the sitter's image, and go on copying it as before. Often, too, the great painter copies better from some conception of his own brain than from actual nature. His mind's eye is surer than his body's. Blake wrote: "Men think they can copy Nature as correctly as I copy imagination. This they will find impossible." And again, "Why are copies of Nature incorrect, while copies of imagination are correct? This is manifest to all." Of course this statement is debatable; but for its philosophy, and for illustrations alike of the definite and the sublime, there is nothing later than Michelangelo to which one refers more profitably than to the life and letters, and to the titanic yet clear and beautiful designs, of the inspired draughtsman William Blake. Did he see his visions? Undeniably. Did he call them into

In the mind's eye.

"This bodiless creation."

absolute existence? Sometimes I think he did; that all soul is endowed with the divine power of creation in the concrete. If so, man will realize it in due time. The poetry of Blake, prophetic and otherwise, must be read with discrimination, for his linguistic execution was less assured than that of his brush and graver; his imagination as a painter, and his art-maxims, were of the high order, but his work as a poet was usually rhapsodical and ill-defined.

But, as I have said, the strength and beauty of any man's poetry depend chiefly upon the definiteness of his mental vision. I once knew a poet of genuine gifts who did not always "beat his music out." When I objected to a feeble, indistinct conception in one of his idyls, "Look you," said he, "I see that just as clearly as you do; it takes hold of me, but I haven't" (he chose to say) "your knack of definite expression." To which I rejoined: "Not so. If you saw it clearly you would express it, for you have a better vocabulary at your command than I possess. Look out of the window, at that building across the street. Now let us sit down, and see who can make the best picture of it in fifteen lines of blank verse — you or I." After a while our trial was completed. His verse, as I had expected, was more faithful and expressive than mine, was apter in word and outline. It reinforced my claim. "There," said I, "if you saw the conception of your other poem as plainly as you see that ordinary building, you would convey it defi-

The conception definite.

nitely. You would not be confused and obscure, for you have the power to express what your mind really pictures."

The true poet, said Joubert, "has a mind full of very clear images, while ours is only filled with confused descriptions." Now, vague- <small>Conceptive faculty of the true poet.</small> ness of impression engenders a kind of excitement in which a neophyte fancies that his gift is particularly active. He mistakes the wish to create for the creative power. Hence much spasmodic poetry, full of rhetoric and ejaculations, sound and empty fury; hence the gasps which indicate that vision and utterance are impeded, the contortions without the inspiration. Hence, also, the "fatal facility," <small>Pseudo-inspiration.</small> the babble of those who write with ease and magnify their office. The impassioned artist also dashes off his work, but his need for absolute expression makes the final execution as difficult as it is noble. Another class, equipped with taste and judgment, but lacking imagination, proffer as a substitute beautiful and recondite materials gathered here and there. Southey's work is an example of this process, and that of the popular and scholarly author of "The Light of Asia" is not free from it; indeed, you see it everywhere in the verse of the minor art-school, and even in Tennyson's and Longfellow's early poems. But the chief vice <small>The turbid shoal.</small> of many writers is obscure expression. Their seeming depth is often mere turbidness, though it is true that thought may be so analytic that its expression must be novel and difficult. Commonplace thought and verse, however clear, certainly are not greater than Browning's, but as a rule the better the poet the more intelligible. There are no stronger conceptions than those of the Book of Job, of Isaiah, Homer, Shakespeare, nor are there any more patent in their simplicity to the common understanding.

Quality, not theme. The imagination in literature is not confined to that which deals with the weird or superhuman. It is true that, for convenience' sake, the selections classed in the best of our anthologies as "Poems of the Imagination" consist wholly of verse relative to nymphs, fairies, sprites, apparitions, and the like. Although this justly includes "Comus" and "The Rime of the Ancient Mariner," there is more fantasy than imagination in other pieces, — in such a piece, for instance, as "The Culprit Fay." No one knows better than the critical editor of "The Household Book of Poetry" that there is more of the high imaginative element in brief touches, such as Wordsworth's

> "The light that never was on sea or land,
> The consecration and the poet's dream, — "

or Shakespeare's

> "Light thickens, and the crow
> Makes wing to the rooky wood, — "

or Bryant's path of the waterfowl, through

> "the desert and illimitable air,
> Lone wandering, but not lost, — "

or Stoddard's vanished city of the waste, —

> "Gone like a wind that blew
> A thousand years ago, — "

and countless other passages as effective, than in the whole of Drake's "Culprit Fay," that being eminently a poem of fancy from beginning to end.

But the imagination is manifold and various. Among its offices, though often not as the most poetic, may be counted invention and construction. These, with characterization, are indeed the chief functions of the novelist. But the epic narratives have been each a growth, not a sudden formation, and the effective plots of the grand dramas — of Shakespeare's, for example — have mostly been found and utilized, rather than newly invented. "The Princess," "Aurora Leigh," and "Lucile" are almost the only successful modern instances of metrical tale-invention, and the last two are really novels in verse. The epic and dramatic poets give imagination play in depicting the event; the former, as Goethe writes to Schiller, conceiving it "as belonging completely to the past," and the latter "as belonging completely to the present." But neither has occasion to originate his story; his concern is with its ideal reconstruction.

Inventive and constructive power.

The imagination, however, is purely creative in the work to which I have just said that it is not restricted, namely, the conception of beings not drawn from experience, to whom it alone can give an existence that is wondrous yet seemingly not out of nature. Such are the forms which Shakespeare called "from the vasty deep": the Weird Sisters, the greenwood sprites, the haunted-island progeny of earth and air. Such are those quite differing creations, Goethe's mocking fiend and the Mephistophilis of Marlowe's "Faustus." Milton's Satan, the grandest of imaginary personages, does not seem to belong to the supramortal class; he is the more sublime because, though scaling heaven and defying the Almighty, he is so unmistakably human. Shakespeare is not strong in the imaginative construction of many of his plays, at least not in the artistic sense, — with respect to that the "Œdipus at Colonos" is a masterpiece, — but he very safely left them to construct themselves. In the conception of human characters, and of their thoughts and feelings, he is still sovereign of imagination's world. In modern times the halls of Wonder have been trodden by Blake and Coleridge and Rossetti. The marvellous "Rime," with its ghostly crew, its spectral seas, its transformation of the elements, is pure and high-sustained imagination. In "Christabel" both the terror and the loveliness are haunting. That beauteous fragment was so potent with the romanticists that Scott formed his lyrical method, that of "The Lay of the Last Minstrel," upon it, and Byron quickly yielded to its spell. But Coleridge's creative mood was as brief as it was enrapturing. From his twenty-sixth to his twenty-eighth year he blazed out like Tycho Brahe's star, then sank his light in metaphysics, exhibiting little thenceforth of worth to literature except a criticism of poets and dramatists that in its way was luminous and constructive.

The poet often conveys a whole picture by a single imaginative touch. A desert scene <small>Suggestiveness.</small> by Gérôme would give us little more than we conceive from Landor's suggestive detail —

> "And hoofless camels in long single line
> Stalk slow, with foreheads level to the sky."

This force of suggestion is nevertheless highly effective in painting, as where the shadow of the cross implies the crucifixion, or where the cloud-phantoms seen by Doré's "Wandering Jew" exhibit it; and as when, in the same artist's designs for Don Quixote, we see visions with the mad knight's eyes. Of a kindred nature is the previ- <small>Prevision.</small> sion, the event forestalled, of a single word or phrase. Leigh Hunt cited the line from Keats' "Isabella," "So the two brothers and their murdered man," — the victim, then journeying with his future slayers, being already dead in their intention. A striking instance of the swift-flashing imagination is in a stanza from Stoddard's Horatian ode upon the funeral of Lincoln: —

> "The time, the place, the stealing shape,
> The coward shot, the swift escape,
> The wife, *the widow's* scream."

What I may call the constant, the *habitual*, imagination of a true poet is shown by his instinct for words,—those keys which all may clatter, and which yield their music to so few. He finds the inevitable word or phrase, unfound before, and it becomes classical in a moment. The power of words and the gift of their selection are uncomprehended by writers who have all trite and hackneyed phrases at the pen's end. The imagination begets original diction, suggestive epithets, verbs implying extended scenes and events, phrases which are a delight and which, as we say, speak volumes, single notes which establish the dominant tone.

<small>Imaginative diction.</small>

This kind of felicity makes an excerpt from Shakespeare unmistakable. Milton's diction rivals that of Æschylus, though nothing can outrank the Grecian's ανήριθμον γέλασμα,— the innumerous laughter of his ocean waves. But recall Milton's "wandering moon" (borrowed, haply, from the Latin), and his "wilderness of sweets;" and such phrases as "dim, religious light," "fatal and perfidious bark," "hide their diminished heads," "the least-erected spirit that fell," "barbaric pearl and gold," "imparadised in one another's arms," "rose like an exhalation," "such sweet compulsion doth in music lie;" and his fancies of the daisies' "quaint enamelled eyes," and of "dancing in the checkered shade;" and numberless similar beauties that we term Miltonic.

<small>"The best words in their best order."</small>

After Shakespeare and Milton, Keats stands first in respect of imaginative diction. His appellatives of the Grecian Urn, "Cold pastoral," and "Thou foster-child of silence and slow time," are in evidence. "The music yearning like a god in pain," and

> " Music's golden tongue
> Flattered to tears this aged man and poor,"

excel even Milton's "forget thyself to marble." What a charm in his "darkling I listen," and his thought of Ruth "in tears amid the alien corn"! Shelley's diction is less sure and eclectic, yet sometimes his expression, like his own skylark, is "an unbodied joy." Byron's imaginative language is more rhetorical, but none will forget his "haunted, holy ground," "Death's prophetic ear," "the quiet of a loving eye" (which is like Wordsworth, and again like Landor's phrase on Milton, — "the Sabbath of his mind"). None would forego "the blue rushing of the arrowy Rhone," or "the dead but sceptred sovereigns, who still rule our spirits from their urns," or such a combination of imagination and feeling as this : —

> "I turned from all she brought to those she could not bring."

Coleridge's "myriad-minded Shakespeare" is enough to show his mastery of words. A conjuring quality like that of the voices heard by Kubla Khan, —

> "Ancestral voices prophesying war, — "

lurks in the imaginative lines of our Southern lyrist,

Boner, upon the cottage at Fordham, which aver of Poe, that

> " Here in the sobbing showers
> Of dark autumnal hours
> He heard suspected powers
>   Shriek through the stormy wood."

Tennyson's words often seem too laboriously and exquisitely chosen. But that was a good moment when, in his early poem of "Œnone," he pictured her as wandering

> "Forlorn of Paris, once her playmate on the hills."

Amongst Americans, Emerson has been the chief master of words and phrases. Who save he could enveil us in "the tumultuous privacy" of the snow-storm? Lowell has great verbal felicity. It was manifest even in the early period when he apostrophized the dandelion, — "Dear common flower," "Thou art my tropics and mine Italy," — and told us of its "harmless gold." But I have cited a sufficient number of these well-wonted instances. Entering the amazing treasure-house of English song, one must remember the fate of the trespasser within the enchanted grotto of the "Gesta Romanorum," where rubies, sapphires, diamonds, lay in flashing heaps on every side. When he essayed to fill his wallet with them, the spell was broken, the arrow whizzed, and he met the doom allotted to pickers and stealers.

With respect to configuration, the antique genius,

in literature as in art, was clear and assured. It imagined plainly, and drew firm outlines. But the Acts and Scenes of our English dramatists were often shapeless; their schemes were full of by-play and plot within plot; in fine, their constructive faculty showed the caprice of rich imaginations that disdained control. Shakespeare, alone of all, never fails to justify Leigh Hunt's maxim that, in treating of the unusual, "one must be true to the supernatural itself." When the French and German romanticists broke loose from the classic unities, they, too, at first went wild. Again, the antique conceptions are as sensuous, beside the modern, as the Olympian hierarchy compared with the spiritual godhood to which Christendom has consecrated its ideals. But whether pagan or Christian, all the supernaturalism of the dark and mystic North has a more awe-inspiring quality than that of sunlit Italy and Greece. There are weird beings in the classic mythology, but its Fates and Furies are less spectral than the Valkyriës and the prophetic Sisters of the blasted heath. Even in the mediæval under-world of Dante, the damned and their tormentors are substantially and materially presented, with a few exceptions, like the lovers of Rimini, — the

> "unhappy pair
> That float in hell's murk air."

Having, then, laid stress upon the excellence of clear

vision, let me add that imaginative genius can force *Compulsion of the Vague.* us to recognize the wonder, terror, and sublimity of the Vague. Through its suggested power we are withdrawn from the firm-set world, and feel what it is

> " to be a mortal
> And seek the things beyond mortality."

What lies beyond, in the *terra incognita* from which we are barred as from the polar spaces guarded by arctic and antarctic barriers, can only be suggested by formlessness, extension, imposing shadow, and phantasmal light. The early Hebraic expression of its mysteries will never be surpassed. Nothing in even the culminating vision of the Apocalypse so takes hold of us as the ancient words of Eliphaz, in the Book of Job, describing the fear that came upon him in the night, when deep sleep falleth on man : —

"Then a spirit passed before my face; the hair of my flesh stood up. It stood still, but I could not discern the form thereof: an image was before mine eyes, there was silence, and I heard a voice, saying: 'Shall mortal man be more just than God? Shall a man be more pure than his Maker?'"

English poetry doubly inherits the sublimity of *Camoëns, Milton, Coleridge.* the vague, from its Oriental and its Gothic strains. Yet it has produced few images more striking than that one which lifts the "Lusiad," by Camoëns, above the level of a perfunctory epic. Vasco da Gama and his crew are struggling to pass the southern point of Africa into the Indian seas beyond. The Spirit of the Cape of Tempests, mantled in blackness of cloud, girt about with lightning and storm, towers skyward from the billows, portentous, awful, vague, and with an unearthly voice of menace warns the voyagers back. I have said that the grandest of English supernatural creations is Milton's Satan. No other personage has at once such magnitude and definiteness of outline as that sublime, defiant archangel, whether in action or in repose. Milton, like Dante, has to do with the unknown world. The Florentine bard soars at last within the effulgence of " the eternal, coeternal beam." Milton's imagination broods " in the wide womb of uncreated night." We enter that " palpable obscure," where there is "no light, but rather darkness visible," and where lurk many a "grisly terror" and "execrable shape." But the genii of wonder and terror are the familiars of a long succession of our English poets. Coleridge, who so had them at his own call, knew well their signs and work; as when he pointed a sure finger to Drayton's etching of the trees which

"As for revenge to heaven each held a withered hand."

Science drives spectre after spectre from its path, but the rule still holds—*omne ignotum pro magnifico*, and a vaster unknown, a more impressive vague, still deepens and looms before.

A peculiarly imaginative sense of the beautiful,

also, is conveyed at times by an exquisite formlessness of outline. I asked the late Mr. Grant White what he thought of a certain picture by Inness, and he replied that it seemed to be "painted by a blind poet." But no Inness, Fuller, Corot, Rousseau, not even Turner, nor the broad, luminous spaces of Homer Martin, ever excelled the magic of the changeful blending conceptions of Shelley, so aptly termed the poet of Cloudland. The feeling of his lyrical passages is all his own. How does it justify itself and so hold us in thrall? Yield to it, and if there is anything sensitive in your mould you are hypnotized, as if in truth gazing heavenward and fixing your eyes upon a beauteous and protean cloud; fascinated by its silvery shapelessness, its depth, its vistas, its iridescence and gloom. Listen, and the cloud is vocal with a music not to be defined. There is no appeal to the intellect; the mind seeks not for a meaning; the cloud floats ever on; the music is changeful, ceaseless, and uncloying. Their plumed invoker has become our type of the pure spirit of song, almost sexless, quite removed at times from earth and the carnal passions. Such a poet could never be a sensualist. "Brave translunary things" are to him the true realities; he is, indeed, a creature of air and light. "The Witch of Atlas," an artistic caprice, is a work of imagination, though as transparent as the moonbeams and as unconscious of warmth and cold. Mary Shelley objected to it on the score that it had no human interest. It certainly is a kind of *aër potabilis*, a wine that lacks body; it violates Goethe's dictum, to wit: "Two things are required of the poet and the artist, that he should rise above reality and yet remain within the sphere of the sensuous." But there is always a law above law for genius, and all things are possible to it — even the entrance to a realm not ordered in life and emotion according to the conditions of this palpable warm planet to which our feet are bound.

As in nature, so in art, that which relatively to ourselves is large and imposing has a corresponding effect upon the mind. Magnitude is not to be disdained as an imaginative factor. An heroic masterpiece of Angelo's has this advantage at the start over some elaborate carving by Cellini. Landor says that "a throne is not built of birds'-nests, nor do a thousand reeds make a trumpet." Of course, if dimension is to be the essential test, we are lost. Every one feels himself to be greater than a mountain, than the ocean, even than Chaos; yet an imaginative observer views the measureless nebula with awe, conceiving a universe of systems, of worlds tenanted by conscious beings, which is to be evolved from that lambent, ambient star-dust. Dimensional effect.

Certain it is that when we seek the other extreme, the province of the microscopic, Fancy, the elf-child of Imagination, sports within her own minute and capricious realm. Her land is that of Fancy.

whims and conceits, of mock associations, of Midsummer Nights' Dreams. She has her own epithets for its denizens, for the "green little vaulter," the "yellow-breeched philosopher," the "animated torrid zone," of her dainty minstrelsy. Poets of imagination are poets of fancy when they choose to be. Hester Prynne was ever attended by her tricksy Pearl. But many is the poet of fancy who never enters the courts of imagination — a joyous faun indeed, and wanting nothing but a soul.

A large utterance, such as that which Keats bestowed upon the early gods, is the instinctive voice of the imagination nobly roused and concerned with an heroic theme. There are few better illustrations of this than the cadences and diction of "Hyperion," a torso equal to the finished work of any other English poet after Shakespeare and Milton; perhaps even greater because a torso, for the construction of its fable is not significant, and when Keats produced his effect, he ended the poem as Coleridge ended "Christabel." All qualities which I have thus far termed imaginative contribute to the majesty of its overture: —

"The grand manner."

> "Deep in the shady sadness of a vale
> Far sunken from the healthy breath of morn,
> Far from the fiery noon, and eve's one star,
> Sat gray-hair'd Saturn, quiet as a stone,
> Still as the silence round about his lair.
> Forest on forest hung about his head
> Like cloud on cloud. No stir of air was there, —
> Not so much life as on a summer's day

> Robs not one light seed from the feather'd grass,
> But where the dead leaf fell, there did it rest.
> A stream went voiceless by, still deaden'd more
> By reason of his fallen divinity
> Spreading a shade: the Naiad 'mid her reeds
> Press'd her cold finger closer to her lips."

At the outset of English poetry, Chaucer's imagination is sane, clear-sighted, wholesome with open-air feeling and truth to life. *Chaucer.* Spenser is the poet's poet chiefly as an artist. The allegory of "The Faerie Queene" is not like that of Dante, forged at white heat, but the symbolism of a courtier and euphuist who felt its unreality. But all in all, the Elizabethan period displays the English imagination at full height. *The Elizabethans.* Marlowe and Webster, for example, give out fitful but imaginative light which at times is of kindred splendor with Shakespeare's steadfast beam. Webster's "Duchess of Malfi" teaches both the triumphs and the dangers of the dramatic fury. The construction runs riot; certain characters are powerfully conceived, others are wild figments of the brain. It is full of most fantastic speech and action; yet the tragedy, the passion, the felicitous language and imagery of various scenes, are nothing less than Shakespearean. To comprehend rightly the good and bad qualities of this play is to have gained a liberal education in poetic criticism.

Now take a collection of English verse, — and there is no poetry more various and inclusive, — take, let us say, Ward's "English Poets," and you will find that the generations after Shakespeare are *Two centuries.* not over-imaginative until you approach the nineteenth century. From Jonson to the Georgian school there is no general efflux of visionary power. The lofty Milton and a few minor lights — Dryden, Collins, Chatterton — shine at intervals between. Precisely the most unimaginative period is that covered by Volume III. and entitled "From Addison to Blake." We have tender feeling and true in Goldsmith and Gray. There is no passion, no illumination, until you reach Burns and his immediate successors. Then imagination leaped again to life, springing chiefly from subjective emotion, as among the Elizabethans it sprang from young adventure, from discovery and renown of arms, above all from the objective study of the types and conduct of mankind. If another century shall add a third imaginative lustre to the poetry of our tongue, — enkindled, perchance, by the flame of a more splendid order of discovery, even now so exalting, — it will have done its equal share.

The Mercury and Iris of this heavenly power are *Comparison, etc.* comparison and association, whose light wings flash unceasingly. Look at Wordsworth's similes. He took from nature her primitive *The elemental bards.* symbolism. Consider his *elemental* quality: I use the word as did the ancients in their large, untutored view of things, — as Prospero uses it, ere laying down his staff: —

> "My Ariel, — chick, —
> That is thy charge: then to the elements
> Be free, and fare thou well!"

In Wordsworth's mind nature is so absolute that her skies and mountains are just as plainly imaged as in the sheen of Derwentwater; and thence they passed into his verse. He wanders, — *Wordsworth.*

> "lonely as a cloud
> That floats on high o'er vales and hills."

He says of Milton : —

> "Thy soul was like a star, and dwelt apart."

A primeval sorrow, a cosmic pain, is in the expression of his dead love's reunion with the elements: —

> "No motion has she now, no force;
> She neither hears nor sees,
> Rolled round in earth's diurnal course,
> With rocks, and stones, and trees."

The souls of the Hebrew bards, inheritors of pastoral memories, ever consorted with the elements, invoking the "heavens of heavens," "the waters that be above the heavens," "fire and hail; snow, and vapor: stormy wind fulfilling His word." Of the Greeks, Æschylus is more elemental than Pindar, even than Homer. Among our moderns, a kindred quality strengthened the imaginations of Byron and Shelley; Swinburne too, whom at his best the Hebraic feeling and the Grecian sway by turns, is most self-forgetful and exalted when giving it full play.

I point you to the fact that some of our American poets, if not conspicuous thus far for dramatic power, have been gifted — as seems fitting in respect to their environment — with a distinct share of this elemental imagination. It is the strength of Bryant's genius: the one secret, if you reflect upon it, of the still abiding fame of that austere and revered minstrel. His soul, too, dwelt apart, but like the mountain-peak that looks over forest, plain, and ocean, and confabulates with winds and clouds. I am not sure but that his elemental feeling is more impressive than Wordsworth's, from its almost preadamite simplicity. It is often said that Bryant's loftiest mood came and went with "Thanatopsis." This was not so; though it seemed at times in abeyance. "The Flood of Years," written sixty-five years later than "Thanatopsis" and when the bard was eighty-two, has the characteristic and an even more sustained majesty of thought and diction.

<small>Bryant. Cp. "Poets of America": pp. 81, 82.</small>

It is easy to comprehend why the father of American song should be held in honor by poets as different as Richard Henry Stoddard and Walt Whitman. These men have possessed one quality in common. Stoddard's random and lighter lyrics are familiar to magazine readers, with whom the larger efforts of a poet are not greatly in demand. But I commend those who care for high and lasting qualities to an acquaintance with his blank verse, and with sustained lyrics like

<small>Stoddard's blank verse, etc.</small>

the odes on Shakespeare and Bryant and Washington, which resemble his blank verse both in artistic perfection and in imagination excelled by no contemporary poet. Whitman's genius is prodigal and often so elemental, whether dwelling upon his types of the American people, or upon nature animate and inanimate in his New World, or upon mysteries of science and the future, that it at times moves one to forego, as passing and inessential, any demur to his matter or manner. There is no gainsaying the power of his imagination,— a faculty which he indulged, having certainly carried out that early determination to loaf, and invite his soul. His highest mood is even more than elemental; it is cosmic. In almost the latest poem of this old bard, addressed "To the Sunset Breeze" (one fancies him sitting, like Borrow's blind gypsy, where he can feel the wind from the heath), he thus expressed it:— *[Whitman's cosmic mood.]*

"I feel the sky, the prairies vast — I feel the mighty northern lakes; I feel the ocean and the forest — *somehow I feel the globe itself swift-swimming in space.*"

Lanier is another of the American poets distinguished by imaginative genius. In his case this became more and more impressible by the sense of elemental nature, and perhaps more subtly alert to the infinite variety within the unities of her primary forms. Mrs. Stoddard's poetry, as yet uncollected, is imaginative and original, the utterance of moods that are only too infre- *[Some other poets.]*

quent. The same may be said of a few poems by Dr. Parsons, from whom we have perhaps the finest of American lyrics, the lines "On a Bust of Dante." There is a nobly elemental strain in Taylor's "Prince Deukalion" and "The Masque of the Gods." I could name several of our younger poets, men and women, and a number of their English compeers, whose work displays imaginative qualities, were it not beyond my province. But many of the new-comers — relatively more, perhaps, than in former divisions of this century — seem restricted to the neat-trimmed playgrounds of fancy and device; they deck themselves like pages, rarely venturing from the palace-close into the stately Forest of Dreams. If one should stray down a gloaming vista, and be aided by the powers therein to chance for once upon some fine conception, I fancy him recoiling from his own imagining as from the shadow of a lion.

On wings above his fate upborne.

Here, then, after the merest glimpse of its aureole, we turn away from the creative imagination: a spirit that attends the poet unbidden, if at all, and compensates him for neglect and sorrow by giving him the freedom of a clime not recked of by the proud and mighty, and a spiritual wealth "beyond the dreams of avarice." Not all the armor and curios and drapery of a Sybaritic studio can make a painter; no æsthetic mummery, no mastery of graceful rhyme and measure, can of themselves furnish forth a poet. Go rather to Barbizon, and see what pathetic truth and beauty dwell within the humble rooms of Millet's cottage; go to Ayr, and find the muse's darling beneath a straw-thatched roof; think what feudal glories came to Chatterton in his garret, what thoughts of fair marble shapes, of casements "innumerable of stains and splendid dyes," lighted up for Keats his borough lodgings. Doré was asked, at the flood-tide of his good fortune, why he did not buy or build a château. "Let my patrons do that," he said. "Why should I, who have no need of it? My château is here, behind my forehead." He who owns the wings of imagination shudders on no height; he is above fate and chance. Its power of vision makes him greater still, for he sees and illuminates every-day life and common things. Its creative gift is divine; and I can well believe the story told of the greatest and still living Victorian poet, that once, in his college days, he looked deep and earnestly into the subaqueous life of a stream near Cambridge, and was heard to say, "What an imagination God has!" Certainly without it was not anything made that was made, either by the Creator, or by those created in his likeness. I say "created," but there are times when we think upon the amazing beauty, the complexity, the power and endurance, of the works of human hands — such as, for example, some of the latest architectural decorations illuminated by the electric light with splendor never conceived of even by an ancestral rhapso-

<small>Creation.</small>

<small>"Ye shall be as Gods."</small>

dist in his dreams of the New Jerusalem — there are moments when results of this sort, suggesting the greater possible results of future artistic and scientific effort, give the theory of divinity as absolutely immanent in man a proud significance. We then comprehend the full purport of the Genesitic record, — "Ye shall be as gods." The words of the Psalmist have a startling verity, — "I have said, Ye are gods; and all of you are children of the Most High." We remember that one who declared himself the direct offspring and very portion of the Unknown Power, and in evidence stood upon his works alone, repeated these words, — by inference recognizing a share of Deity within each child of earth. The share allotted to such a mould as Shakespeare's evoked Hartley Coleridge's declaration : —

> "The soul of man is larger than the sky,
> Deeper than ocean — or the abysmal dark
> Of the unfathomed centre. . . .
> So in the compass of the single mind
> The seeds and pregnant forms in essence lie
> That make all worlds."

But what was the old notion of the act of divine creation? That which reduced divinity to the sprite of folk-lore, who by a word, a spell, or the wave of a wand, evoked a city, a person, an army, out of the void. The Deity whom we adore in our generation has taken us into his workshop. We see that he creates, as we construct,

The infinite process.

slowly and patiently, through ages and by evolution, one step leading to the next. I reassert, then, that "as far as the poet, the artist, is creative, he becomes a sharer of the divine imagination and power, and even of the divine responsibility." And I now find this assertion so well supported, that I cannot forbear quoting from "A Midsummer Meditation" in a recent volume of American poetry: — *"Two Worlds." By R. W. Gilder.*

> "Brave conqueror of dull mortality!
> Look up and be a part of all thou see'st; —
> Ocean and earth and miracle of sky,
> All that thou see'st thou art, and without thee
> Were nothing. Thou, a god, dost recreate
> The whole; breathing thy soul on all, till all
> Is one wide world made perfect at thy touch.
> And know that thou, who darest a world create,
> Art one with the Almighty, son to sire —
> Of his eternity a quenchless spark."

We have seen that with the poet imagination is the essential key to expression. The other thing of most worth is that which moves him to expression, the passion of his heart and soul. I close, therefore, by saying that without either of these elements we can have poetry which may seem to you tender, animating, enjoyable, and of value in its way, but without imagination there can be no poetry which is great. Possibly we can have great poetry which is devoid of passion, but great only through its tranquillizing power, through tones that calm and strengthen, yet do not exalt and thrill. *Incentive.*

Such is not the poetry which stirs one to make an avowal like Sir Philip Sidney's:—

"I never heard the old song of Percy and Douglas, that I found not my heart moved more than with a trumpet."

VIII.

THE FACULTY DIVINE.

POETIC expression is that of light from a star, our straightest message from the inaccessi- *Unde aether sidera pascit?* ble human soul. Critics may apply their —*Lucr.* spectral analysis to the beam, but without such a process our sympathetic instinct tells us how fine, how rude, how rare or common, are the primal constituents from which its vibrations are derived. The heat-rays, the light, the actinic,— these may be combined in ever various proportions, but to make a vivid expression they must in some proportion come together. Behind the action at their starting-place glows and pulsates a spirit of mysterious and immortal force, the "vital spark," to comprehend which were to lay hold upon divinity itself. As to the poet's share of this, Wordsworth, that inspired schoolmaster with the gift to create a soul under the ribs of pedantry, conceived his impressive title,— "the faculty divine." Before approaching more closely to this radiant source, we have to touch upon one remaining element which seems most of all to excite its activity, and to which, in truth, a whole discourse might be devoted as equitably as to truth, or beauty, or imagination.

I have laid stress, heretofore, upon the passion which so vivifies all true poetry that certain thinkers believe the art has no other office than to give emotion vent. And I have just said that, while poetry which is not imaginative cannot be great, the utterance which lacks passion is seldom imaginative. It may tranquillize, but it seldom exalts and thrills. Therefore, what is this quality which we recognize as passion in imaginative literature? What does Milton signify, in his masterly tractate on education, by the element of poetry which, as we have seen, he mentions last, as if to emphasize it? Poetry, he says, is simple, — and so is all art at its best; it is sensuous, — and thus related to our mortal perceptions; lastly, it is passionate, — and this, I think, it must be to be genuine.

<small>Passion. See pages 19, 49.</small>

In popular usage the word "passion" is almost a synonym for love, and we hear of "poets of passion," votaries of Eros or Anteros, as the case may be. Love has a fair claim to its title of the master passion, despite the arguments made in behalf of friendship and ambition respectively, and whether supremacy over human conduct, or its service to the artistic imagination, be the less. Almost every narrative-poem, novel, or drama, whatsoever other threads its coil may carry, seems to have love for a central strand. Love has the heart of youth in it,

<small>Not an epithet of love alone.</small>

> " — and the heart
> Giveth grace unto every art."

Love, we know, has brought about historic wars and treaties, has founded dynasties, made and unmade chiefs and cabinets, inspired men to great deeds or lured them to evil: in our own day has led more than one of its subjects to imperil the liberty of a nation, if not to deem, with Dryden's royal pair, "the world well lost." A strenuous passion indeed, and one the force of which pervades imaginative literature.

But if Milton had used the word "impassioned," his meaning would be plainer to the vulgar apprehension. Poetic passion is intensity of emotion. *Passion and Imagination.* Absolute sincerity banishes artifice, ensures earnest and natural expression; then beauty comes without effort, and the imaginative note is heard. We have the increased stress of breath, the tone, and volume, that sway the listener. You cannot fire his imagination, you cannot rouse your own, in quite cold blood. Profound emotion seems, also, to find the aptest word, the strongest utterance, — not the most voluble or spasmodic, — and to be content with it. Wordsworth speaks of "thoughts that do often lie too deep for tears," while Mill says that "the poetry of a poet is Feeling itself, using thought only as a means of expression." The truth is that passion uses the imagination to supply conceptions for its language. On the other hand, the poet, imagining situations and experiences, becomes excited through dwelling on them. But whether passion or imagination be first aroused, they speed together like the wind-sired horses of Achilles.

The mere artisan in verse, however adroit, will do well to keep within his liberties. Sometimes you find one affecting the impassioned tone. It is a dangerous test. His wings usually melt in the heat of the flame he would approach. Passion has a finer art than that of the æsthete with whom beauty is the sole end. Sappho illustrated this, even among the Greeks, with whom art and passion were one. Keats felt that "the excellence of every art is its intensity, capable of making all disagreeables evaporate, from their being in close relations with beauty and truth." Passion rises above the sensuous, certainly above the merely sensual, or it has no staying power. I heard a wit say of a certain painting that it was "repulsive equally to the artist, the moralist, and the voluptuary." Even in love there must be something ideal, or it is soon outlawed of art. A few of Swinburne's early lyrics, usually classed as erotic, with all their rhythmic beauty, are not impassioned. His true genius, his sacred rage, break forth in measures burning with devotion to art, to knowledge, or to liberty. There is more real passion in one of the resonant "Songs before Sunrise" than in all the studiously erotic verse of the period, his own included.

Emotion must be unaffected and ideal.

The idea that poetry is uttered emotion, though now somewhat in abeyance, is on the whole modern. It was distinctive with the romantic school,

until the successors of Scott and Byron allied a
new and refined tenderness to beauty. *Recognition of this force in art.*
The first rush had been that of splendid
barbarians. It is so true that strong natures recognize the force of passion, that even Wordsworth, conscious of great moods, was led to confess that "poetry is the spontaneous outflow of powerful feelings," and saved himself by adding that it takes "its origin from emotion recollected in tranquillity." Poets do retain the impressions of rare moments, and express them at their own time. But "the passion of Wordsworth," under which title *Wordsworth's emotional limits.* I have read an ingenious plea for it by Dr. Coan, was at its best very serene, and not of a kind to hasten dangerously his heart-beats. Like Goethe, he regarded human nature from without; furthermore, he studied by choice a single class of people, whose sensibilities were not so acute, say what you will, as those of persons wonted to varied and dramatic experiences. The highest passion of his song was inspired by inanimate nature; it was a tide of exaltation and worship, the yearning of a strong spirit to be at one with the elements. Add to this his occasional notes of feeling: the pathos of love in his thought of Lucy:—

> "But she is in her grave, and, oh,
> The difference to me!"

the pathos of broken comradeship in the quatrain:—

> "Like clouds that rake the mountain-summits,
> Or waves that own no curbing hand,

> How fast has brother followed brother
> From sunshine to the sunless land!"

include also his elevated religious and patriotic moods, and we have Wordsworth's none too frequent episodes of intense expression.

All passion obtains relief by rhythmic utterance in music or speech; it is soothed like Saul in his frenzy by the minstrel harp of David. But the emotion which most usually gives life to poetry is not that of fits of passion, but, as in the verses just quoted, of the universal moods embraced in the word "feeling." Out of natural feeling, one touch of which "makes the whole world kin," come the lyrics and popular verse of all nations; it is the fountain of spontaneous song. Take the poetry of this class from Southern literatures, such as the Italian and Spanish, and you leave only their masterpieces. At first thought, it seems more passionate than our own, and certainly it is more sonorous. But Anglo-Saxon words are deep and strong, although there is a good deal of insularity in the song from "The Princess":—

[sidenote: The quality of Feeling.]

> "O tell her, Swallow, thou that knowest each,
> That bright and fierce and fickle is the South,
> And dark and true and tender is the North."

If this be so, they should wed indissolubly, for each must be the other's complement. Scottish verse is full of sentiment, often with the added force of pathos. For pure feeling we all

[sidenote: Voices of the heart.]

carry in our hearts "Auld Lang Syne," "The Land o' the Leal," Motherwell's "Jeanie Morrison," and " My heid is like to rend, Willie." Robert Burns is first and always the poet of natural emotion, and his fame is a steadfast lesson to minstrels that if they wish their fellow-men to feel for and with them, they must themselves have feeling. Only from the depths of a great soul could come the stanzas of "Highland Mary" and "To Mary in Heaven." He touches chords for high and low alike in the unsurpassable "Farewell":—

> "Had we never loved sae kindly,
> Had we never loved sae blindly,
> Never met or never parted,
> We had ne'er been broken-hearted!"

His lyrics of joy, ambition, patriotism, are all virile with the feeling of a brave and strong nature.

English emotional verse is more self-conscious, and often flooded with sentimentalism. Yet Byron's fame rests upon his intensity, whether that of magnificent apostrophes, or of his personal poems, among which none is more genuine than his last lyric, written upon completing his thirty-sixth year. In the Victorian period the regard for art has covered sentiment with an aristocratic reserve, but Hood was a poet of emotion in his beautiful songs and ballads no less than in "The Bridge of Sighs." *English sentiment.*

From the middle register of emotion, poetry rises

to the supreme, such as that of Shelley's "Lines to an Indian Air," or the more spiritual ecstasy of his invocation to the West Wind:

The ecstasy of song.

> "Make me thy lyre, even as the forest is:
> What if my leaves are falling like its own!
> The tumult of its mighty harmonies
> Will take from both a deep autumnal tone,
> Sweet though in sadness. Be thou, spirit fierce,
> My spirit! Be thou me, impetuous one!"

Of recent English lyrical poets Mrs. Browning is one of the most impassioned. Her lips were touched with fire; her songs were magnetic with sympathy, ardor, consecration. But our women poets of the century usually have written from the heart; none more so than Emma Lazarus, whose early verse had been that of an art-pupil, and who died young,— but not before she seized the harp of Judah and made it give out strains that all too briefly renewed the ancient fervor and inspiration.

"Das Ewig-weibliche."

Every note of emotion has its varying organ-stops: religious feeling, for instance, whether perfectly allied with music in cloistral hymns, or expressed objectively in studies like Tennyson's "St. Agnes" and "Sir Galahad," and Elizabeth Lloyd's "Milton in his Blindness," or rising to the eloquent height of Coleridge's Chamouni Hymn. So it is with martial songs and national hymns, from Motherwell's "Cavalier's Song," and Campbell's "Ye Mariners of England," to the Marseillaise hymn, to "My Maryland"

"Fill all the stops of Life with tuneful breath."

and the "Battle Hymn of the Republic." It is the passion of Lowell's "Memorial Odes" that so lifts their rhythm and argument. With Poe, beauty was a passion, but always hovering with strange light above some haunted tomb. Emerson exhibits the intensity of joy as he listens to nature's "perfect rune." On the one side we have Poe avowing that the "tone" of the highest manifestations of beauty is one of sadness. "Beauty of whatever kind," he said, "in its supreme development, invariably excites the sensitive soul to tears." This is the key-note of our romanticism, of which there has been no more sensitive exemplar than Poe, — Grecian as he was at times in his sense of form. But far more Grecian, in temper and philosophy, was Emerson, who found the poet's royal trait to be his cheerfulness, without which "no man can be a poet, for beauty is his aim. . . . Beauty, the spirit of joy and hilarity, he sheds upon the universe." What diverse interpretations, each a lesson to those who would limit the uncharted range of feeling and art! Yet it is easy to comprehend what Poe meant, and to confess that mortal joy is less intense of expression than mortal grief. And it was Emerson himself who, in his one outburst of sorrow, gave us the most impassioned of American lyrics, the "Threnody" for his lost child, — his "hyacinthine boy." This free and noble poem — even for its structural beauty, so uncommon in Emerson's work — must rank with

[sidenote: Exquisite sadness and enraptured joy.]

[sidenote: Emerson's "Threnody."]

memorable odes. But the poet's faith, thought, imagination, are all quickened by his sorrow, so that the "Threnody" is one of the most consolatory as well as melodiously ideal elegies in the language.

Taken for all in all, Whittier, "our bard and prophet best-beloved," that purely American minstrel, so virginal and so impassioned, at once the man of peace and the poet militant, is the Sir Galahad of American song. He has read the hearts of his own people, and chanted their emotions, and powerfully affected their convictions. His lyrics of freedom and reform, in his own justified language, were "words wrung from the nation's heart, forged at white heat." Longfellow's national poems, with all their finish, cannot rival the natural art of Whittier's ; they lack the glow, the earnestness, the intense characterization, of such pieces as " Randolph of Roanoke," "Ichabod," and "The Lost Occasion." The Quaker bard, besides, no less than Longfellow, is a poet of sympathy. Human feeling, derived from real life and environment, is the charm of "Snow-Bound," even more than its absolute transcript of nature. Years enough have passed since it was written for us to see that, within its range, it is not inferior to "The Deserted Village," "The Cotter's Saturday Night," and "Tam O'Shanter."

Mark Pattison justly declared that "poets of the first order" always have felt that "human action or

Whittier's impulsive glow. Cp. "Poets of America": pp. 121-128.

passion" is the highest theme. These are the topics of Homer, Dante, Milton, Goethe, Hugo. Dante, while perceiving by the smiling of the stars, and by the increasing beauty and divineness of Beatrice, that she is translating him to the highest spheres, still clings to his love for the woman. Its blood-red strand connects his Paradise with earth. The Faust-Margaret legend is human to the radiant end. Rossetti's "The Blessed Damozel" idealizes the naïve materialism of the cathedral ages. The motive of that prismatic ballad is the deathless human passion of the sainted maiden. Her arms make warm the bar of Heaven on which she leans, still mortal in her immortality, waiting for the soul of her lover. Such is the poetic instinct that no creature can be finer in quality, however advanced in power, than man himself; that the emotions of his soul are of the uttermost account. Rossetti was ever an impassioned poet, in whom were blended Northern and Italian types. His series of sonnets, "The House of Life," quivers with feeling. Christina Rossetti, his sister, holds her eminence not by the variety and extent of her verse, but for its emotion deep inwrought. Tennyson's career indicates that the line of advance for a poet is that of greater intensity; nevertheless, he has furnished a typical example of the national repugnance to throwing wide the gates of that deep-set but rugged castle, an English heart. His sense

Art's highest theme.

The human element.

From youth to age. Cp. "Victorian Poets": p. 422.

of beauty and art at first was all in all, although such poems as "Locksley Hall" and "The Sisters" — such a line as that from the former, —

"And our spirits rushed together at the touching of the lips," —

showed him capable of taking up the "Harp of Life." Throughout his long idyllic reign, he grew upon the whole more impassioned in thought and dramatic conception, — yet the proof of this is not found in his dramas, but in portions of "In Memoriam," in powerful studies like "Lucretius" and "Rizpah," and in the second part of "Locksley Hall." Great poets confront essentials as they approach their earthly resolution.

<small>The objective creation of impassioned types.</small> Thus far I have referred only to the emotion of the poet's own soul, often the more intense and specific from its limits of range. The creative masters give us all the hues of life's "dome of many-colored glass," as caught from their interior points of view. What is life but the speech and action of us all, under stress of countless motives and always of that blind emotion which Schopenhauer termed the World-Will? It is at the beck of the strong invoker that these modes of feeling come arrayed for action, and not in single spies, but far more various than the passions which Collins's Muse drew around her cell. Such are the throes of Homer's personages within and without the walls of Troy. The intense and natural emo-

tion of Priam and Achilles, of Hector and Andromache and Helen, has made them imperishable. The heroic epics have gone with their ages, and for every romantic and narrative poem we have a hundred novels; but the drama remains, with its range for the display of passion's extreme types. The keen satisfaction we take in an exhibi- See page 103. tion, not of the joy and triumph alone, but of the tragedy, the crime, the failure of lives that ape our own, is not morbid, but elevating. We know by instinct that they are right who declare all passion good *per se;* we feel that it is a good servant if a bad master, and bad only when it goes awry, and that the exhibition of its force both enhances and instructs the force within each soul of us. Again, the poet who broods on human passion Exaltation. and its consequent action attains his highest creative power: he rises, as we say, at each outbreak and crisis, and the actor impersonating his conception must rise accordingly, or disappoint the audience which knows that such culminations are his opportunities, above the realistic level of a well-conceived play. More than all, and as I have suggested in a former lecture, the soul looks tran- Intense sensations enhance the worth of life. quilly on, knowing that it, no more than its prototypes, can be harmed by any mischance. "Agonies" are merely "its changes of garments." They are forms of *experience*. The soul desires *all* experiences; to touch this planetary life at all points, to drink not of triumph and delight alone; it needs must drain its portion of anguish, failure, wrong. It would set, like the nightingale, its breast against the thorn. Its greatest victory is when it is most agonized. When all is lost, when the dark tower is reached, then Childe Roland dauntless winds his blast upon the slug-horn. Its arms scattered, its armor torn away, the soul, "the victor-victim," slips from mortal encumbrance and soars freer than ever. *Victor atque victima, atque ideo victor quia victima.* This is the constant lesson of the lyrics and plays and studies of Browning, the most red-blooded and impassioned of modern dramatic poets; a wise and great master, whose imagination, if it be less strenuous than his insight and feeling, was yet sufficient to derive from history and experience more types of human passion than have been marshalled by any compeer. I have been struck by a critic's quotation of a passage from Beyle (written in 1817) which says that, after centuries of artificiality, it must be the office of the coming artist to express "states of soul," — that that is what a Michelangelo would do with modern sculpture. In truth the potent artist, the great poet, is he who makes us realize the emotions of those who experience august extremes of fortune. For what can be of more value than intense and memorable sensations? What else make up that history which alone is worth the name of life?

The most dramatic effects are often those which indicate suppressed passion — that the hounds are ready to slip the leash. These are constantly utilized by Browning; they characterize the Puritan repression in Hawthorne's romances and Mrs. Stoddard's novels, and the weird power of Emily Brontë's "Wuthering Heights." In the drama, above all, none but a robustious periwigpated fellow is expected to "tear a passion to tatters." Nor can dramatic heights be of frequent occurrence: they must rise like mountains from a plain to produce their effect, and even then be capped with clouds — must have something left untold. A poem at concert-pitch from first to last is ineffective. See with what relief of commonplace or humor Shakespeare sets off his supreme crises: the banter with Osric before the death of Hamlet; the potter and babble of the peasant who brings the aspic to Cleopatra. In the silent arts, as in nature, the prevailing mood is equable, and must be caught. The picture on your walls that displays nature in her ordinary mien, and not in a vehement and exceptional phase, is the one which does not weary you. But poetry, with its time-extension, has the freedom of dramatic contrasts — of tranquillity and passion according to nature's own allotment. With this brave advantage, naturalism is ignoble which restricts itself to the ordinary, and is indeed grossly untrue to our life, at times so concentrated and electric. *[margin: Reserved power.]* *[margin: True naturalism.]*

The ideal of dramatic intensity — that is, of *imagined* feeling — is reached when the expression is as inevitable as that of a poet's outburst under stress of personal emotion. You are conscious, for example, that one must endure a loss as irreparable as that which Cowper bemoaned, before he can realize the pathos and beauty of the monody " On the Receipt of my Mother's Picture " : —

Absolute dramatic passion.

> "O that those lips had language! Life has passed
> With me but roughly since I heard thee last."

But you also feel, and as strongly, that only one who has been agonized by the final surrender, whether to violence or death, of an adored child, can fully comprehend that passionate wail of Constance bereft of Arthur : —

> "Grief fills the room up of my absent child,
> Lies in his bed, walks up and down with me,
> Puts on his pretty looks, repeats his words,
> Remembers me of all his gracious parts,
> Stuffs out his vacant garments with his form;
> Then have I reason to be fond of grief."

Shakespeare's dramas hold the stage, and if his stronger characters are not impersonated so frequently as of old, they are still the chief rôles of great actors, and are supported with a fitness of detail unattained before. The grand drama, then, is the most efficient form of poetry in an unideal period to conserve a taste for something imaginative and impassioned. But, with a public bred to reserve, our new plays and poems on the whole avoid extremes of feeling, which, alike in life

Modern equanimity.

and literature, are not "good form." What we do accept is society drama, chiefly that which turns upon the Parisian notion of life as it is. But whether the current drama, poetic or otherwise, reflects life as it is, is a question upon which I do not enter. I have referred to the lack of passion in modern poetry. The minor emotions are charmingly, if lightly, expressed. Humor, for instance, is given a play almost Catullian; and that Mirth is a feeling, if not a passion, is the lyrical justification of some of our felicitous modern song. Many of our poets realize that we have rounded a beautiful but too prolonged idyllic period; they amuse themselves with idly touching the strings, while awaiting some new dispensation — the stimulus of a motive, the example of a leader. Emotion cannot be always sustained; there must be intervals of rest. But each generation desires to be moved, to be thrilled; and they are mistaken who conceive the poetic imagination to be out of date and minstrelsy a foible of the past.

As it is, we hear much talk, on the part of those observers whose business it is to record the movement of a single day, about the decline of ideality. *An idle outcry.* Whenever one of the elder luminaries goes out, the cry is raised, Who will there be to take his place? What lights will be left when the constellation of which he was a star shall have vanished? The same cry has gone up from every generation in all eras. Those who utter it are like water-beetles perceiving only the ripples, comprehending little of the great waves of thought and expression, upon which we are borne along. The truth is that, alike in savagery and civilization, there never is a change from stagnation to life, from bondage to freedom, from apathy to feeling and passion, that does not beget its poets. At such a period we have the making of new names in song, as surely as deeds and fame in great wars come to men unknown before. It is true that the greatest compositions, in all the arts, are usually produced at culminating epochs of national development. But the period of that eminent group, the "elder American poets," surely has not been that of our full development. Theirs has been the first inspiring rise of the foot-hills, above which — after a stretch of mesa, or even a slight descent — range upon range are still to rise before we reach that culminating sierra-top whose height none yet can measure. Throughout this mountain-climbing, every time that a glowing and original poet appears, his art will be in vogue again.

Now, is such a poet the child of his period, or does he come as if by warrant and create an environment for himself? From the first it seemed to me a flaw in the armor of Taine, otherwise our most catholic exponent of the principles of art, that he did not allow for the irrepressibleness of genius, for the historic evidence that now and then "God lets loose a man in the world." Such

When comes the poet.

a man, it is true, must be of ingrained power to overcome an adverse situation; his very originality will for a long time, as in the recent cases of Wordsworth and Browning, stand in his way, even if in the end it secures for him a far more exceeding crown of glory. If the situation is ripe for him, then his course is smooth, his work is instantly recognizable. First, then, the poet is needed. He must possess, besides imaginative and emotional endowments, the special gifts which, however cultivable, come only at birth — "the vision and the faculty divine," and a certain strong compulsion to their exercise. But these gifts, under such compulsion, constitute what we mean by the poet's genius. *The Faculty Divine.*

In our age of distributed culture, it has become a matter of doubt — even among men reared upon the Shorter Catechism — whether there is any predestination and foreordination of the elect in art, literature, or action. Many deem this a superstition which has too long prevailed. That it has impressed mankind everywhere and always is a matter of record. I have much faith in a universal instinct; and I believe that I still have with me the majority even of modern realists, and that the majority is right, in refusing to discredit the gift of high and exceptional qualities to individuals predestined by heredity or otherwise, and I believe that without this gift — traditionally called genius — no poet has afforded *Genius: whether it is " the inspired gift of God."*

notable delight and service. I know that men of genius often waive their claim; that Buffon said genius was "but long-continued patience"; that Carlyle wrote, it "means transcendent capacity for taking trouble, first of all"; that one eminent modern writer, though in a passing mood, announced: "There is no 'genius'; there is only the mastery which comes to natural aptitude from the hardest study of any art or science." But these are the surmises of men whose most original work comes from them so easily that they do not recognize the value of the gift that makes it natural. They honestly lay more stress upon the merit of the hard labor which genius unconsciously drives them to undertake. I say "drives them," and call to mind Lowell's acute distinction: "Talent is that which is in a man's power; genius is that in whose power a man is." Carlyle's whole career proves that he simply wished to recognize the office laid upon genius of taking "infinite trouble." His prevailing tone is unmistakable: "Genius," he says, "is the inspired gift of God." "It is the clearer presence of God Most High in a man;" and again, "Genius, Poet, do we know what those words mean? An inspired Soul once more vouchsafed to us, direct from Nature's own fire-heat, to see the Truth, and speak it, or do it." His whole philosophy of sway by divine right is a genius-worship. Even Mr. Howells's phrase, "natural aptitude," if raised to the highest power, is a recognition of something behind mere industry. It is what forces the hero, the artist, the poet, to be absorbed in a special office, and decides his choice of it.[1]

The world is equipped with steadfast workers whose natural taste and courageous, strenuous labor do not lift them quite above the mediocre. The difference between these, the serviceable rank and file, and the originative leaders, is one of kind, not of degree. However admirable their skill and service in time become, they do not get far apart from impressions common to us all. We cannot dispense with their army in executive and mechanical fields of action. It is a question whether they are so essential to arts of taste and investigation ; to philosophy, painting, music ; to the creative arts of the novelist and poet. But with respect to these, it would be most unjust to confound them with the upstarts whose condign sup-

Talent and executive service.

---

[1] Nothing of late has seemed apter than a criticism of the *Saturday Review* upon certain outgivings of the academicians, Sir Frederick Leighton and Sir John Millais, quite in the line of the industrial theory from which the present writer is dissenting. The reviewer, commenting upon these didactic paradoxes, asserts that all the truth which is in them amounts to just this: "That the intuitive perceptions and rapidity of combination which constitute genius, whether in action or speculation, in scientific discovery or inventive art or imaginative creation, open out so many new problems and ideas as to involve in their adjustment and development the most arduous labor and the most unwearied patience. But without the primal perception the labor will be vanity and the patience akin to despair. Perhaps it is important to keep in mind that labor without the appropriate capacity is even more fruitless than aptitude without industry."

pression is a desirable thing for both the public and themselves,—claimants really possessed of less than ordinary sense. Such is the fool of the family who sets up for a "genius"; the weakling of the borough, incapable of practical work, or too lazy to follow it, but with a fondness for fine things and a knack of imitating them. Such are the gadflies of every art, pertinaciously forcing themselves upon attention, and lowering their assumed crafts in the esteem of a community.

<small>Pretension.</small>

It is wise to discriminate, also, between genius and natural fineness of taste. The latter, joined with equally natural ambition, has made many a life unhappy that had peculiar opportunities for delight. For surely it is a precious thing to discern and enjoy the beautiful. Taste in art, in selection, in conduct, is the charm that makes for true aristocracy, a gift unspoiled but rather advanced by gentle breeding, a grace in man, and adorable in woman; it is something to rest content with, the happier inasmuch as you add to the happiness of others. It is the nimbus of many a household, beautifying the speech and bearing of the members, who, if they are wise, realize that their chief compensation is a more tranquil study and possession of the beautiful than the fates allot to those who create it. Hephaistos, the grim, sooty, halt artificer of all things fair, found small comfort even in the possession of Aphrodite, the goddess who inspired him. The secret of happiness, for a refined

<small>Taste, as distinguished from artistic genius.</small>

nature, is a just measure of limitations. Taste is not always original, creative. There are no more pathetic lives than the lives of those who know and love the beautiful, and who surrender its enjoyment in a vain struggle to produce it. Their failures react upon finely sensitive natures, and often end in sadness, even misanthropy, and disillusionment when the best of life is over.

Men of talent and experience do learn to concentrate their powers on certain occasions, and surprise us with strokes like those of genius. That is where they write "better than they can," as our Autocrat so cleverly has put it. But such efforts are exhausting and briefly sustained. I know it is said that genius also expires when its work is done; but who is to measure its reservoir of force, or to gauge the unseen current which replenishes it? *Fortunate moments.*

That there is something which comes without effort, yet impels its possessor to heroic labor, is immemorially verified.[1] It whispered melodies to Mozart almost in his boyhood, made him a composer at five, — at seven the author of an opus, four sonatas for piano and violin ; and it so drew him on to victorious industry, that he asserted in after life: "No one has taken such pains with the study of composition as I!" It made the *Congenital gift.*

---

[1] The cases of Mozart and Dickens, with others equally notable, were cited by the writer in an extended paper on Genius, which was published several years ago.

child Clairon, as she refused to learn to sew, cry out under brutal punishment: "Kill me! You had better do so, for if you don't I shall be an actress!" Dickens declared that he did not invent his work: "I see it," he said, "and write it down."[1] Sidney Lanier, in nervous crises, would seem to hear rich music. It was an inherited gift. Thus equipped with a rhythmical sense beyond that of other poets, he turned to poetry as to the supreme art. Now, the finer and more complex the gift, the longer exercise is needful for its full mastery. He strove to make poetry do what painting has done better, and to make it do what only music hitherto has done. If he could have lived three lives, he would have adjusted the relations of these arts as far as possible to his own satisfaction. I regard his work, striking as it is, as merely tentative from his own point of view. It was as if a discoverer should sail far enough to meet the floating rock-weed, the strayed birds, the changed skies, that betoken land ahead; should even catch a breath of fragrance wafted from out-

---

[1] Hartmann's scientific definition, which I cited in a former lecture, — "Genius is the activity and efflux of the intellect freed from the domination of the conscious Will," — finds its counterpart in the statement by F. W. H. Myers, concerning the action of the "Subliminal Consciousness." This, Mr. Myers says, has to do with "the initiation and control of organic processes, which the conscious will cannot reach. . . . Perhaps we seldom give the name of genius to any piece of work into which some uprush of subliminal faculty has not entered." (See the *Journal of the Society for Psychical Research*, February, 1892.)

lying isles, and then find his bark sinking in the waves before he could have sight of the promised continent.

In our day, when talent is so highly skilled and industry so habitual, people detect the genius of a poet or tale-writer through its originality, perhaps first of all. It has a different note, even in the formative and imitative period, and it soon has a different message,— perhaps one from a new field. The note is its style; the message involves an exhibition of creative power. Genius does not borrow its main conceptions. As I have said, it reveals a more or less populous world of which it is the maker and showman. Here it rises above taste, furnishing new conditions, to the study of which taste may profitably apply itself. It is neither passion nor imagination, but it takes on the one and makes a language of the other. Genius of the universal kind is never greater than in imparting the highest interest to good and ordinary and admirable characters; while a limited faculty can design only vicious or eccentric personages effectively, depending on their dramatic villainy or their grotesqueness for a hold upon our interest. Véron has pointed out this inferiority of Balzac and Dickens to Shakespeare and Molière—and he might have added, to Thackeray also. In another way the genius of many poets is limited, — that of Rossetti, of Poe, for example,— poets of few, though striking, tones and of isolated temperaments. Genius of the

more universal type is marked by a sound and healthy
Sanity. judgment. You may dismiss with small respect the notion of Fairfield, Lombroso, and their like, that genius is the symptom of neurotic disorder — that all who exhibit it are more or less mad. This generalization involves a misconception of the term; they apply it to the abnormal excess, the morbid action, of a special faculty, while true genius consists in the creative gift of one or more faculties at the highest, sustained by the sane coöperation of the possessor's other physical and mental endow-
Wisdom. ments. Again, what we term common sense is the genius of man as a race, the best of sense because the least ratiocinative. Nearly every man has thus a spark of genius in the conduct of life. A just balance between instinct, or understanding, and reason, or intellectual method, is true wisdom. It requires years for a man of constructive talent — a writer who forms his plans in advance — for such a man to learn to be flexible, to be obedient to his sudden intuitions and to modify his design
Obedience to the vision. accordingly. You will usually do well to follow a clew that comes to you in the heat of work — in fact, to lay aside for the moment the part which you had designed to complete at once, and to lay hold of the new matter before that escapes you. The old oracle, Follow thy
Spontaneity. genius, holds good in every walk of life. Everything, then, goes to show that genius is that force of the soul which works at its own seemingly

capricious will and season, and without conscious effort; that its utterances declare what is learned by spiritual and involuntary discovery: —

> "Vainly, O burning Poets!
> Ye wait for his inspiration,
> Even as kings of old
> Stood by Apollo's gates.
> *Hasten back*, he will say, *hasten back
> To your provinces far away!
> There, at my own good time,
> Will I send my answer to you.*"

Yes, the spontaneity of conception, which alone gives worth to poetry, is a kind of revelation — the imagery of what genius perceives by Insight. This sense has little to do with reason and induction; it is the inward light of the Quaker, the *a priori* guess of the scientist, the prophetic vision of the poet, the mystic, the seer. If it be direct vision, it should be incontrovertible. In occult tradition the higher angels, types of absolute spirit, were thought to know all things by this pure illumination: —

Revelation through Insight.

> "There, on bright hovering wings that tire
> Never, they rest them mute,
> Nor of far journeys have desire,
> Nor of the deathless fruit;
> For in and through each angel soul
> All waves of life and knowledge roll,
> Even as to nadir streams the fire
> Of their torches resolute."

While this is a bit of Preraphaelite mosaic, it is not too much to say of the essentially poetic soul that at times it becomes, in Henry More's language, —

> "One orb of sense, all eye, all airy ear;"

that it seems to have bathed, like Ayesha, in central and eternal flame; or, after some preëxistence, to have undergone the lustration to which, in the sixth Æneid, we find the beclouded spirits subjected: —

> "Donec longa dies, perfecto temporis orbe,
> Concretam exemit labem, purumque relinquit
> Aetherium sensum atque aurai simplicis ignem." [1]

At such times its conclusions are as much more infallible than those worked out by logic as is the offhand pistol-shot of the expert, whose weapon has become a part of his hand, than the sight taken along the barrel. It makes the leopard's leap, without reflection and without miss. I think it was Leigh Hunt who pointed out that feeling rarely makes the blunders which thought makes. Applied to life, we know that woman's intuition is often wiser than man's wit.

The clearness of the poet's or artist's vision is so much beyond his skill to reproduce it, and so increases with each advance, that he never quite contents himself with his work:

The artist's noble discontent.

---

[1] "Till Time's great cycle of long years complete
Clears the fixed taint, and leaves the ethereal sense
Pure, a bright flame of unmixed heavenly air."
Cranch's Translation.

Hence the ceaseless unrest and dissatisfaction of the best workman. His ideal is constantly out of reach, — a "lithe, perpetual escape."

From the poet's inadequate attempts at expression countless myths and faulty statements have originated. Still, he keeps in the van of discovery, and has been prophetic in almost every kind of knowledge, — evolution not excepted, — and from time immemorial in affairs that constitute history. This gave rise, from the first, to a belief in the direct inspiration of genius. Insight de- *Inspiration.* rives, indeed, the force of inspiration from the sense that a mandate of utterance is laid upon it. To the ancients this seemed the audible command of deity. "The word of the Lord came unto me, saying," — "Thus saith the Lord unto me," — "So the spirit lifted me up and took me away, and I went in bitterness, in the heat of my spirit, but the hand of the Lord was strong upon me," — such were the avowals of one of the greatest poets of all time. The vision of Ezekiel and the compulsion to declare it have been the inspiration of *The prophetic* the prophetic bard, of the impassioned *gift.* lyric poet, almost to our own day. His time has passed. We cannot have, we do not need, another Ezekiel, another Dante or Milton. Hugo, the last Vates, was the most self-conscious, and his own deity. A vision of the wisdom and beauty of art has inspired much of the superior poetry of recent times. A few prophetic utterances have been heard, evoked in some struggle of humanity, some battle for liberty of belief or nationality or conduct. Yet I doubt not that, whenever a great cause is in progress, — before its culminating triumph, rather than after, — it will have its impassioned and heroic minstrelsy. The occasion will seek out and inspire its poet.

But he must believe in his prophecy, and as *Indispensability of Faith.* something greater than himself, though indomitably believing in himself as the one appointed to declare it. Reflecting upon the lack of originality, of power, of what we may consider tokens of inspiration, in so much of our most beautiful latter-day song, I suspect that it is not due alone to the diversion of effort in many new fields of action and expression, but also to a general doubt of the force and import of this chief art of expression, — even to the modern poet's own distrust in its significance. The higher his gift and training, the more he seems affected by the pleasant cynicism which renders him afraid, above all, of taking himself and his craft "too seriously." This phrase itself is the kind of chaff which he most dreads to incur. Now, I have just spoken of the wisdom of recognizing one's limitations, but if one has proved that he has a rare poetic gift, I think that he scarcely can take it and himself too seriously. The poets of our language and time who have gained the most distinction — such as Tennyson, Browning,

Longfellow, Arnold, Emerson — have taken themselves very seriously indeed; have refused to go after strange gods, and have done little but to make poetry or to consider matters demanding the higher exercise of thought and ideality. Doubtless poets are born nowadays as heretofore, though nature out of her "fifty seeds" may elect to bring not even "one to bear." But some who exhibit the most command of their art, and in truth a genuine faculty, are very shy of venturing beyond the grace and humor and tenderness of holiday song.

I think that such a condition might be expected to exist during the unsettled stage of con- *Cynicism.* viction now affecting our purpose and imagination. There is no lack of desire for a motive, but an honest lack of motive, — a questioning whether anything is worth while, — a vague envy, perhaps, of the superb optimism of our scientific brethren, to whom the material world is unveiling its splendors as never before, and to whom, as they progress so steadfastly, everything seems worth while.

I remember an impressive lyric, perhaps the finest thing by a certain American writer. Its title, "What is the Use?" was also the burden of his song. He took his own refrain so much to heart that, although he still lives according to its philosophy, there are only a few of us who pay meet honor to him as a poet.

Distinction ever has been achieved through some form of faith, and even the lesser poets have won their respective measures of success, other things being equal, in proportion to their amount of trust in certain convictions as to their art, themselves, and "the use of it all." The serene forms of faith in deity, justice, nationality, religion, human nature, which have characterized men of the highest rank, are familiar to you. Such faiths have been an instinct with sovereign natures, from the Hebraic sense of a sublime Presence to the polemic belief of Bunyan and Milton. Homer cheerfully recognizes the high gods as the inspirers and regulators of all human action. Dante's faith in the ultimate union of perfect beauty and perfect holiness was intense, and his conviction in the doom of the ignoble was so absolute that he felt himself commissioned to pronounce and execute it. Shakespeare made no question of the divinity that doth hedge a king; he believed in institutions, in sovereignty, in the English race. His tranquil acceptance of the existing order of things had no later parallel until the century of Goethe and Emerson and Browning. Byron and Shelley invoked political and religious liberty, and believed in their own crusade against Philistia. Hugo and his band were leaders in a lifelong cause; they carried a banner with "Death to tradition" upon it. The underlying motive of all strenuous and enthusiastic movement, in art or poetry, is faith. Gautier and Musset concerned themselves with beauty and romantic passion; Clough and Arnold, with philosophy and feel-

[sidenote: Faith of some kind the stay of all true art.]

ing: all were poets and knights-errant according to their respective tempers and nationalities. And so we might go on indefinitely, without invalidating the statement that some kind of faith, with its resulting purpose, has engendered all poetry that is noteworthy for beauty or power. True art, of every class, thrives in an affirmative and motive-breeding atmosphere. It is not the product of cynicism, pessimism, or hopeless doubt. I do not mean "the honest doubt" which Tennyson sets above "half the creeds." The insatiate quest for light is nobler than a satisfied possession of the light we have. The scientific unsettlement of tradition is building up a faith that we are obtaining a new revelation, or, at least, opening our eyes to a continuous one.

But without surmising what stimulants to imaginative expression may be afforded hereafter, let me refer to a single illustration of the creative faith of the poet. *A crowning masterpiece of faith.* For centuries all that was great in the art and poetry of Christendom grew out of that faith. What seems to me its most poetic, as well as most enduring, written product, is not, as you might suppose, the masterpiece of a single mind,—the "Divina Commedia," for instance, —but the outcome of centuries, the expression of many human souls, even of various peoples and races. Upon its literary and constructive side, I regard the venerable Liturgy of the historic Christian Church as one of the few world-poems, the poems universal. *The Church Liturgy.* I care not which of its rituals you follow, the Oriental, the Alexandrian, the Latin, or the Anglican. The latter, that of an Episcopal Prayer-Book, is a version familiar to you of what seems to me the most wonderful symphonic idealization of human faith,—certainly the most inclusive, blending in harmonic succession all the cries and longings and laudations of the universal human heart invoking a paternal Creator.

I am not considering here this Liturgy as divine, Its universal quality. though much of it is derived from what multitudes accept for revelation. I have in mind its human quality; the mystic tide of human hope, imagination, prayer, sorrows, and passionate expression, upon which it bears the worshipper along, and wherewith it has sustained men's souls with conceptions of deity and immortality, throughout hundreds, yes, thousands, of undoubting years. The Orient and Occident have enriched it with their finest and strongest utterances, have worked it over and over, have stricken from it what was against the consistency of its import and beauty. It has been A growth. a growth, an exhalation, an apocalyptic cloud arisen "with the prayers of the saints" from climes of the Hebrew, the Greek, the Roman, the Goth, to spread in time over half the world. It is The voice of human brotherhood. the voice of human brotherhood, the blended voice of rich and poor, old and young, the wise and the simple, the statesman and the clown; the brotherhood of an age which, knowing little, comprehending little, could have no refuge save trust in the oracles through which a just and merciful Protector, a pervading Spirit, a living Mediator and Consoler, had been revealed. This being its nature, and as the crowning masterpiece of faith, you find that in various and constructive beauty — as a work of poetic art — it is unparalleled. It is lyrical from first to last with perfect and melodious forms of human speech. Its chants and anthems, its songs of praise and hope and sorrow, have allied to themselves impressive music from the originative and immemorial past, and the enthralling strains of its inheritors. Its prayers are not only for all sorts and conditions of men, but for every stress of life which mankind must feel in common — in the household, or isolated, or in tribal and national effort, and in calamity and repentance and thanksgiving. Its wisdom is forever old and perpetually new; its calendar celebrates all seasons of the rolling year; its narrative is of the simplest, the most pathetic, the most rapturous, and most ennobling life the world has known. There is no malefactor so wretched, no just man so perfect, as not to find his hope, his consolation, his lesson, in this poem of poems. I have called it lyrical; it is dramatic in structure and effect; it is an epic of the age of faith; but in fact, as a piece of inclusive literature, it has no counterpart, and can have no successor. Time and again some organization for worship and instruction, building its foundations upon reason rather than on faith, has

Its symphonic perfection.

Without a parallel.

tried to form some ritual of which it felt the need. But such a poem of earth and heaven is not to be made deliberately. The sincere agnostic must be content with his not inglorious isolation; he must barter the rapture and beauty and hope of such a liturgy for *his* faith in something different, something compensatory, perchance a future and still more world-wide brotherhood of men.

Until this new faith, or some fresh interpretation of past belief, becomes vital in action, becomes more operative, the highest flight of poetry will be timidly essayed. The songs of those who are crying, "They have taken away my Lord, and I know not where they have laid him!" will be little else than tenebræ — cries out of the darkness, impassioned, it may be, but hardly forceful or creative.

<small>Tenebræ.</small>

<small>Arnold and Clough.</small> I have spoken of Arnold and Clough, the conspicuously honest, noble, intellectual poets of the transition period. Just as far as their faith extended, their verse rests firmly in art and beauty, love, and nobility of purpose. But much of it comes from troubled hearts; its limits are indicated by a spirit of unrest — limits which Arnold was too sure and fine a self-critic not to perceive; so that, after he had reached them, — which was not until he had given us enduring verse, and shown how elevated was his gift, — he ceased to sing, and set himself resolutely to face the causes of his unrest, and to hasten, through his prose investigations,

the movement toward some new dawn of knowledge-brightened faith.

A few verses from his "Dover Beach" are in the key of several of his most touching lyrics, — in the varying measure so peculiarly his own, — utterances of a feeling which in the end seems to have led him to forego his career as a poet: "The sea of faith," he plains, — *The troubled heart.*

> "Was once, too, at the full, and round earth's shore
> Lay like the folds of a bright girdle furl'd.
> But now I only hear
> Its melancholy, long, withdrawing roar,
> Retreating, to the breath
> Of the night-wind, down the vast edges drear
> And naked shingles of the world.
>
> "Ah, love, let us be true
> To one another! for the world, which seems
> To lie before us like a land of dreams,
> So various, so beautiful, so new,
> Hath really neither joy, nor love, nor light,
> Nor certitude, nor peace, nor help for pain;
> And we are here as on a darkling plain
> Swept with confused alarms of struggle and flight,
> Where ignorant armies clash by night."

Doubtless Arnold's reserve intensified this sadness. Clough equally felt the perturbed spirit of his time; but he had a refuge in a bracing zest for life and nature, which so often made the world seem good to him, and not designed for naught.

In time our poets will acquire, with the new learning and the more humane and critical theology, the health and optimism in which a note- *The new day.*

worthy art must originate if at all. As for the new learning —

> "Say, has the iris of the murmuring shell
> A charm the less because we know full well
> Sweet Nature's trick? Is Music's dying fall
> Less finely blent with strains antiphonal
> Because within a harp's quick vibratings
> We count the tremor of the spirit's wings?
> There is a path by Science yet untrod
> Where with closed eyes we walk to find out God.
> Still, still, the unattained ideal lures,
> The spell evades, the splendor yet endures;
> False sang the poet, — there is no good in rest,
> And Truth still leads us to a deeper quest."

For one, I believe that the best age of imaginative production is not past; that poetry is to retain, as of old, its literary import, and from time to time to prove itself a force in national life; that the Concord optimist and poet was sane in declaring that "the arts, as we know them, are but initial," that "sooner or later that which is now life shall be poetry, and every fair and manly trait shall add a richer strain to the song."

Thoughts in conclusion. And now, after all that has been said in our consideration of the nature of poetry, and although this has been restricted closely to its primal elements, I am sensible of having merely touched upon an inexhaustible theme; that my comments have been only "words along the way." Meanwhile the press teems with criticism, our time is alert with debate in countless private and public assemblies respecting almost every verse of all renowned poets, ancient or contemporary; texts and editions, even if relatively less in number compared with the varied mass of publications, are multiplied as never before, and readers — say what you may — are tenfold as many as in the prime of the elder American minstrels. The study of poetry has stimulated other literary researches. Yet the best thing that I or any one can say to you under these conditions is that a breath of true poetry is worth a breeze of comment; that one must in the end make his own acquaintance with its examples and form his judgment of them. Read the best; not the imitations of imitations. Each of you will find that with which he himself is most in touch, and therewith a motive and a legend — *petere altiora*. The poet's verse is more than all the learned scholia upon it. He makes it by direct warrant; he produces, and we stand by and often too complacently measure his productions. In no wise can I forget that we are regarding even the lowliest poets from our still lower station; we are like earth-dwellers viewing, comparing, mapping out the stars. Whatsoever their shortcomings, their gift is their own; they bring music and delight and inspiration. A singer may fail in this or that, but when he dies the charm of his distinctive voice is gone forever.

ANALYTICAL INDEX

ANALYTICAL INDEX

- ACADEMIC, THE, revolts against, and rise of new schools, 148, 150, 151; cause of its despotism, 157; value of its standards, 159; Sir Joshua Reynolds, 161, 162.
Action, of the drama, 105; defended by Arnold, 133; as the poet's theme, 268.
Actor, the, 271.
Adam Bede, Mrs. Cross, 137.
Addison, 101, 250.
"Adonais," Shelley, 90, 124.
Æneid, The, Vergil, 91–93, 212, 286.
Æschylus and the Greek drama, 98, 99; and see 46, 169, 240, 251.
Æstheticism, less artistic than emotion, 262.
Æsthetics, Poe on Beauty and Taste, 26; Berkeleian theory of, 148, 149; Véron, in his *L'Æsthetique*, 152, 157; and see *Beauty* and *Taste*.
Affectation of feeling, 262.
"Agincourt," Drayton, 94.
Agnosticism, the sincere, 294.
Alastor, Shelley, 124.
Alcæus, 87.
Alcestis, Euripides, 99.
Alexandrian Library, 168.

Alexandrian Period, the Sicilian style, 89, 90.
Alfieri, 128, 133.
Allegory, of Dante, Spencer, and Bunyan, 114; and see 249.
America, theory of her institutions, 3; American quality should pervade our native poetry and sculpture, 200; now on trial, 229.
American Poetry, Longfellow and his mission, 91; its fidelity to Nature, 195; its "elemental" feeling, 252–254; Whittier and Longfellow, 268; the "elder American poets," 276, 297; and see 225, 242.
American School. See *American Poetry*.
Amiel, 135; quoted, 196.
Anacreon, 93.
Analytic Poetry, Browning, 108; Browning's method compared with Shakespeare's, etc., 191, 192; and see 80.
Ancient Mariner, The Rime of the, Coleridge, 81, 125.
Anna Karénina, Tolstoi, 137.
Anthology, the Greek, 88, 169, 183; the Latin, 92.
Anthropomorphism, the **artists'**

true conception of deity, 222, 223.
Antique, the, classical conception of poetry and the poet, 17-19; illustrated by Guido's Aurora, 29, by Homer's "Vision," *ib.*; comprehension of nature's rythm, 52; sculpture, 63; ancient classification of poetry, 76; spirit of an Athenian audience, 79; classicism of Keats and Landor, 124; in modern Italian poetry, 128; Arnold's early subjection to, 133, 134; Schlegel on, 134; our compensation for its loss, 139, 143; the Academic, 157; perfection of, 159; its simplicity, 175, 176; expression of its own time, 199; informing yet objective view of nature, 207, 208; English "classical" style, 213; genius for configuration, 242; the pagan supernaturalism, 243; unison of passion and art, 262; Emerson's philosophy, 267; and see *Hellenism*.
Arabian Nights, The, Galland's, Payne's and Burton's translations, 82; and see 193.
Architecture, served by the other arts, 64; Japanese, La Farge on, 163.
Ariosto, 112.
Aristophanes, and the drama, 99; and see 79, 88, 190.
Aristotle, his view of the nature of poetry, 17-19; relations to Plato, 21; and see 27.
Arnim, 118.

Arnold, E., 82, 235.
Arnold, M., as Goethe's pupil, 19; poetry as a criticism of life, 27, 28; "Thyrsis," 90; conflict of his critical theory with his own genius, 133-135; preface to his second edition, 133, and poems conforming to, 133, 134; subjective lyrics, 134; temperament and career, 135; his selections from Wordsworth, 172; on the Wordsworthians, 219; his beauteous unrest, 294; quoted, 118, 194, 295; and see 162, 218, 289, 290.
"Ars Victrix," Dobson, quoted, 173.
Art, substructural laws of, 6, 7; consensus and differentiation of its modes, 50; it must have life, 70; "Art for Art's sake," 129, 167; its beauteous paradox, 181; not artifice, 201; Goethe and Haydon, *ib.*; has a truth of its own, 202; cause of our delight in, *ib.*; vice nullifies the force of, 216; its absolute liberty, 220; the artist's labor a natural piety, 221; artistic nonconformity, 222; the artist's God, 222, 223; God the master-artist, *ib.*; clearness and retentive faculty of the musician and painter, 232-234, — of the poet, 234, 235; heightened by passion, 262; must express states of soul, 272; repose and true naturalism, 273; and modern inspiration, 287; its best atmosphere, 291; and see *Artistic*

Quality, *The Fine Arts*, *Composite Art*, etc.
Arte of English Poesie, The, Puttenham, 198.
Artificiality, 48, 177.
Artisanship, 226.
Artistic Dissatisfaction, 286.
Artistic Quality, heightened by passion, 128; extreme recent finish, 129, 130; often in excess of originality, 131; Swinburne's, 132.
Art Life, the, studio and table talk, 12; and see 221–223.
Art School, the, recent characteristics of, 130, 131; of the Nineteenth Century, 173; the minor, 235.
Arts, the fine, their practical value, 14; consensus of, 15; music, painting, etc., as compared with poetry, 63–72; Lessing's canon, 66; the "speechless" arts normally objective, 80; must express the beautiful, 147; illustrative of poetry, 155; a Japanese at the Art Students' League, 165; and see *Architecture, Music, Painting, Sculpture*.
Aryan Literature, 87.
Association, 250.
"Astrophel," Spenser, 90.
Atalanta in Calydon, Swinburne, 132.
Auerbach, B., novelist, 137.
"Auld Lang Syne," Burns, 264.
"Auld Robin Gray," Lady Barnard, 194.
Aurora Leigh, Mrs. Browning, 237.

Ausonius, 169.
Austen, Jane, novelist, 138.

BACON, on Poetry, 23; quoted, 203; and see 57.
Balder Dead, Arnold, 134, 135.
Ballads, early English, 94; Thackeray's, 215; and see 194.
Balzac, quoted, 34; and see 137, 283.
Banville, Th. de, 35, 131, 158.
Barnard, Lady, quoted, 194.
Bascom, J., critic, 20.
"Battle Hymn of the Republic," Mrs. Howe, 267.
Baudelaire, 133.
Beauty, proclaimed the sole end of Poetry, by Schlegel and Poe, 26; poetry as an expression of, 46–48; its arbiter, Taste, 47; false standards of, 48; *considered as an element in poetry and cognate arts*, 147–185; its expression an indispensable function, 147; recurrent denials of its indispensability, 147 *et seq.*; these are merely revolts against hackneyed standards, 148, 150, 151, 158; its immortal changefulness, 148; whether it is a chimera, 148, 152–158; this theory purely Berkeleian, and repulsive to the artistic instinct, 149, 155; the "transcendental" contempt for, 149; Emerson's recognition of, 149; impressionism merely a fresh search for, 150; exists in some guise in every lasting work of art, 151; the new Æsthetics,

as set forth by E. Véron, 152, 153; its truth and its fallacies, *ib. et seq.*; derives specific character from its maker's individuality, 152, 153; what B. really is, *viz.*, a quality regulating the vibratory expression of substances, 153-155; all impressions of it unite in spiritual feeling, 154; moral and physical analogous, 154; perception of it is subjective, 155; its quality objective, *ib.*; its connection with the perfection of nature and the fitness of things, 156, with utility, *ib.*; the natural quality of all things, 156; recognized intuitively by the poet, 157; unconsciously postulated even by Véron, 157; danger of irreverence for, 158; national and racial ideals of, 159-165; the Grecian, 159; of the Renaissance, 160; zest for, associated with novelty, *ib.*; the English academic standard, 161; antipodal conceptions of, the Japanese, 162-165; specific evolution of, 164; not fully transferable by translation, etc., 166; essential to the endurance of a poem, or other work of art, 166-173; considered here in the concrete, 167; symbolizes Truth in pure form, 168; the poet's instinct for, *ib.*; has conserved the choicest part of ancient poetry, *ib.*; its effect in the poetry of our own tongue, 170-172; present revival of love for, 172, 173; its concrete poetic elements, 173-180; of melody and scriptural effect, 174; of construction, 174; of simplicity, 175; of variety, 176; of naturalness, 176; of decoration and detail, 177; psychical, 177, 178; of the pure lyric, 178-180; of Charm, 179, 180; of the suggestion of Evanescence, 181-185; what is meant by its unity with Truth, 187, 188, 220-224; and sadness, 267; and see 75; also *Æsthetics, Taste*, etc.

Berkeleianism, as applied to poetry and the arts, 15.

Berkeley, his ideal philosophy, 149, 155.

Beyle, M. H. (Stendhal), quoted, 272.

Bible, Poetry of the, 82-87; the Hebraic genius, 83; racial exaltation of, 83, 84; intense personal feeling of the Psalmists, etc., 84, 85; naïveté and universality, 85; Book of Job, 86; Esther and Ruth, 87, 175; sublimity of, 244; its " elemental " quality, 251; and see 55, 191, 194.

Bion, 90.

"Bishop Blougram's Apology," Browning, 109.

Bizarre, The, 158.

Blackmore, novelist, 157.

Blake, W., quoted, 233; genius of, *ib.*; and see 58, 158, 238, 250.

Blank Verse, English, the noblest dramatic measure, 105.

Bleak House, Dickens, 137.
"Blessed Damozel, The," Rossetti, 269.
Blot in the 'Scutcheon, A, Browning, 110.
Boccaccio, 57, 101.
Boileau, 18.
Boner, J. H., quoted, 242.
Bonnat, L. J. F., artist, 9.
Brahmanism. See *Orientalism*.
Breadth, a mark of superiority in portraying life and nature, 191.
Bride of Lammermoor, The, Scott, 137.
"Bridge of Sighs, The," Hood, 265.
Bridges, R., 179; song by, quoted, 185.
Bronté, Charlotte, 137.
Bronté, Emily, 137, 273.
Browning, his use of rhyme, 56; the master of analytic and psychological drama, 108–110; individuality of, 108; how his work is subjective, 109; method of, *ib.*; Swinburne on, *ib.*; dramatic lyrics and monologues, *ib.*; compared with Shakespeare, 109, 110; and the stage, 110; his statement of the poet's art, 24; as a critical idealist, 169; as a dramatist, 191, 192; his nature-touches, 192, 193; as a thinker and moralist, 213; his estimate of Shelley, 219; his types of passion, 272; originality of, 277; quoted, 197; and see, 35, 42, 60, 69, 136, 142, 215, 288, 290.
Browning, Mrs., passion and beauty of her self-expression, 128; compared with Sappho, 88; "Aurora Leigh," 237; and see 4, 177, 266.
Bryant, W. C., his broad manner, 194, 195; elemental mood of, 252; quoted, 237; and see 129, 210.
Bucolic verse. See *Greek Bucolic Poets, Nature, Idyllic Poets*, etc.
Buddhism. See *Orientalism*.
Buffon, on Genius, 278.
Bull, Lucy C., a child's recognition of Poetry, 124.
Bülow, H. von, musician, 232.
Bunyan, 52, 290.
Burns, his spontaneity, 120; quoted, 265; and see 135, 172, 173, 190, 195, 250, 255.
Burroughs, J., 62.
Burton, R. F., translator, 82.
Butcher, S. H., translator, 82.
Byron, chief of the English Romantic School, 19; view of Poetry, *ib.*; estimate of Pope, *ib.*; considered, 120–123; the typical subjective poet, 121; his "Childe Harold," 121; voice of his period, 122, 123; "Don Juan," 123; compared with Shelley, 124; compared with Heine, 125, 126; his unrest as a poet of nature, 203; influenced by Coleridge, 238, 239; imaginative language of, 241; quoted, 206, 244; and see 58, 60, 119, 142, 173, 195, 226, 251, 263, 290.

CABANEL, A., painter, 9.

Calderon, 79, 100, 101.
Callimachus, Elegiacs on Heracleitus, 89.
Camoëns, 79, 101, 112, 244.
Campbell, 266.
Canning, G., 94.
Carlyle, on inspiration, 23, 24; cited, 196; and see 58.
Carr, J. W. C., cited, 68.
Catholicity, 220.
Catullus, 92, 155, 169.
Cavalier Poets, 168.
"Cavalier's Song," Motherwell, 266.
Cellini, B., artist, 167, 247.
Cenci, The, Shelley, 69, 124.
Cervantes, 79, 101, 191.
Chapman, G., on poetry, 18.
Characterization, dramatic, 105; the novelist, 237.
Charm, of the perfect lyric, 179–182; of Evanescence, 181, 185.
Chatterton, 250, 255.
Chaucer, as a poet of the beautiful, 170; his imagination, 249; and see 115, 131, 215.
"Chevy Chase," 258.
Childe Harold, Byron, 121.
"Childe Roland," Browning, 109, 272.
"Children in the Wood, The," quoted, 194.
Chinese literature, 81.
Christabel, Coleridge, 125, 238, 248.
Christendom, Poetry of, its characteristics, 79; transfer of the Oriental spirit, 82; its epic masterpieces, 112–118; its poetry of Faith, 291; and see 243.

Christendom, The Muse of. See *A. Dürer.*
Christianity, contrasted with Paganism, 139–143; effect of its introspection and sympathy upon poetry, 139 *et seq.*; "The Muse of Christendom," 140, 141; its sublime seriousness, 143; and see 112 *et seq.*
Church, Christian, the mediæval, 140.
Cicero, believer in inspiration, 22.
Citation of Shakespeare, The, Landor, 125.
Clairon, actress, 282.
Classicism, in France, 18, 120; tribute to Prof. Gildersleeve, 100; pseudo, 144, 199, 200; of Queen Anne's time, 213; modern, 225; and see *The Academic* and *The Antique.*
Classification of the poetic Orders, 76.
Clearness of the artistic vision, 232.
Cleopatra, Haggard and Lang, 89.
Cloister and the Hearth, The, Reade, 137.
Clouds, The, Aristophanes, 190.
Clough, A. H., and Arnold's "Thyrsis," 90; his unrest, 294; zest of, 295; and see 135, 290.
Coan, T. M., on "the passion of Wordsworth," 263.
Coleridge, H., quoted, 256.
Coleridge, S. T., his instinct for beauty, 20; concord with Wordsworth as to poetry, imagination, science, 20, 28; genius of, 125; how moved by Nature, 202;

"Ancient Mariner," 236; height and decline of his imagination, 238, 239; mastery of words, 241; cited, 81; quoted, 101; and see 27, 58, 119, 142, 162, 170, 173, 226, 245, 248, 266.
Collins, Mortimer, quoted, 97.
Collins, W., 142, 172, 184, 250, 270.
"Colonial" Revival, the recent, 160, 161.
Comedy, Aristophanes and Molière, 99, 100; Terence and Plautus, 100.
Commonplace, its use as a foil, 273; and see *Didacticism*.
Common Sense, 284.
Comparison, 250.
Complexity, undue, 175.
Composite Art, 76, 164.
Composite Period, the new, 136.
Comus, Milton, 236.
Conception, spontaneity of, 285; and see *Imagination*.
Configuration, of outline, imagery, etc., 242.
Conscious Thought, 147.
Consensus of the Arts, 50, 163, 199.
Construction, poetic architecture, 174; of plot, etc., 174-176; may be decorated, 177; of a sustained work, 178; imaginative, 237; of the Elizabethan dramatists, 243; and see 4.
Contemplative Poetry, Wordsworth's, 219; and see *Truth*, *Didacticism*, etc.
Cook, A. S., editor of Sidney's *Defense*, 23.
Cooper, J. F., novelist, 137.

Corot, painter, 246.
"Correctness" in Art, 162.
Cory's paraphrase on Callimachus, 89; "Mimnermus," 183.
Cotter's Saturday Night, The, Burns, 268.
Counterpoint, musical, 64.
Couture, painter, 10; quoted, 142.
Cowper, 142, 173, 190, 214; quoted, 274.
Cranch, C. P., quoted, 286.
Creative Eras. See *Objectivity*.
Creative Faculty, shared by the artist with his Maker, 44, 45; the modern, exercised upon prose fiction, 137, 138; secret of its genius, 163; Shakespeare's pure, 230; its divinity, 234; of the pure imagination, 237; as godlike, 254-258; men as gods, 256; imagined types of passion, 270; and see *Objectivity*.
Creative Quality. See *Objectivity*.
Criticism, "applied," distinguished from pure, 8; poets as critics, 12; not iconoclastic, 16; recent analytic study of poetry by the public, 41; with respect to modern culture, 59, 60; by law, 80; sometimes to be deprecated, 96; the present a good time for poetic, 138; the best, 142; with respect to beauty, 147 *et seq.*; English critics of poetry, 162; Browning's, 170; false estimates of Shelley, 218; faulty, 219; Coleridge's, 239; on "states of soul," 272; a recurring question, 275; the present an age of, 296, 297; and see 249, 259.

Culprit Fay, The, Drake, 236, 237.
Culture, its success and limitations in art, 61; and see 277.
Cynicism, 289.

DANTE, minor works of, 79; characterized, 112–115; one with his age and poem, 112; intense personality of, 113; Parsons' Lines on a Bust of, 114; compared with Milton, 115, 117, 245; his supernaturalism, 243; human feeling, 269; faith of, 290; quoted, 174; and see 76, 101, 217, 249, 287.
Davenant, Sir W., quoted, 197.
Davies, John, on music, 65.
"Day Dream, The," Tennyson, 68–70.
Decamerone, Il, Boccaccio, 131.
Decoration, 4; and see *Technique.*
Defence of Poetry, A, Shelley, 25.
Defense of Poesie, The, Sidney, 23.
Definiteness of the artistic imagination, 232, 234.
Definition of Poetry, why evaded, 12–15; is possible, 15–17; and see *Poetry.*
De Foe, 138.
Delaroche, H. [Paul], painter, 10.
De l'Isle, Rouget, *La Marseillaise,* 266.
Deluge, The, Sienkiewicz, 137.
De Quincey, 58.
Derby, Lord, 82.
De Rerum Natura, Lucretius, 212, 217.
Descriptive Poetry, word-painting of Spenser, Keats, Tennyson, etc., 67–70; landscape a background to life, 177; not very satisfactory, 202; inferior to painting, *ib.*; when subjective, 202–204; and see 189, 195, 196, also POETRY and *Nature.*
Deserted Village, The, Goldsmith, 268.
Detail. See *Technique.*
Dickens, his prose and verse, 57; quoted, 282; and see, 137, 283.
Diction, of the past, 34; Hugo's, 120; English, 215; the modern vocabulary, 225; imaginative mastery of words, epithets, phrases, 240–242; verbal felicity of Shakespeare, Milton, Keats, Byron, etc., 240, 241,— of Emerson, 242; Fancy's epithets, 248; majestic utterance in "Hyperion," etc., 248; and see *Language.*
Didacticism, of minor transcendentalists, 24; Coleridge's metaphysical decline, 125; the "didactic heresy," why opposed to true poesy, 187, 188, 213; the "higher" and philosophical, 211–213; poetry of wisdom, 211; Ecclesiastes, *ib.;* the Greek sages, Lucretius, Epicurus, Omar, Tennyson, etc., 212, 213; Pope as a moralist-poet, 213–215; of the commonplace, 219; and see *Truth, Ethics,* etc.
Dilettanteism, 8, 133.
Dimension, effect of magnitude in art, 247.
Dimitri Rudini, Tourgénieff, 137.

Directness. See *Style*.
Divina Commedia, Dante, 112–115; charged with its author's personality, 113; symbolism of, 114; compared with "Paradise Lost," 115; and see 269, 291.
Dobson, Austin, 94, 158.
Don Juan, Byron, 19, 123.
Donoghue, J., sculptor, 13, 200.
Don Quixote, Cervantes, 239.
"Dora," Tennyson, 193.
Doré, G., painter, 239; quoted, 255.
Doubt. See *Faith*.
"Dover Beach," Arnold, quoted, 295.
Drake, J. R., 236, 237.
Drama, The. Grand drama the supreme poetic structure, 105–107; analysis of *The Tempest*, 106; impersonality of the masters, 107; modern and subjective, of Browning, 108–110; Browning's genius and method, 108; the modern stage, 110; adaptation to the stage, *ib.*; Jonson on the stage, *ib.*; Swinburne's plays, 132; meretricious plays, 216; the grand drama again, 274; modern plays, society-drama, etc., 274, 275; and see *Elizabethan Period*, *Greek Dramatists*, etc.
Dramatic Lyrics, Browning, 109.
Dramatic Poetry, its narrative may well be borrowed, 57, 237; Shakespeare, Browning, Keats, Shelley, 69, 191, 243; text of, 191; Elizabethan dramatists, 75; epical drama of Job, 86; youthful poems of dramatists, 101; Aristotle on Tragedy, 103; why tragedy elevates the soul, 103, 104, 271, 272; Greek recognition of Destiny, 104; dramatists of Christendom, *ib.*; the dramatic genius, *ib.*; Shakespeare and impersonality, 104, 105; *Faust*, 119; Shelley's, 124; truth to nature, 189, 190, — to life, 191; the Attic, 191; Webster's *Duchess of Malfi*, 249; display of passion's extreme types, 271–273; exaltation of, 271; Browning's types of passion, 272; effect of contrasts, 273.
Dramatic Quality, of Browning's lyrics, etc., 109; evinced of late in prose fiction rather than in poetry, 137, 138; and see *The Drama*.
Dramatists. See *The Drama* and *the Greek Dramatists*.
Dramatists, the Greek, 97–100; ethical motive of, 97; their objectivity, *ib.*; the Attic stage, 99; grand drama as an imaginative transcript of life, 101–104; impersonal, *ib.*; follow the impartial law of nature, 102; and see 113.
Drayton, 94; quoted, 245.
Dryden, Aristotelian view of poetry, 18; quoted, 261; and see 162, 172, 250.
Du Bellay, 171.
Dumas, Père, 138.
Duran, C., painter, 9.

Dürer, A., artist, his "Melencolia" as the Muse of Christendom, 140, 141.
"Dying Christian, The," Pope, 214.

"EACH AND ALL," Emerson, 220, 221.
Earthly Paradise, The, Morris, 131.
Eccentricity, 109.
Ecclesiastes, 211.
Edda, The, 131.
Edison, T., inventor, 32.
Education, the higher and ideal, 4.
Egoism, the Parnassian, 80; and see 140, also *Subjectivity.*
Elaboration, undue, 193.
Elegiac Poetry, Grecian epitaphs, the anthologies, etc., 88, 89; the Greek idyllists, 89, 90; English elegies, 90; Ovid, 92; Latin feeling, 92, 93; Emerson's "Threnody," 267.
"Elemental" Quality, 250–254; Wordsworth's, 251; of the Hebrews, Greeks, and modern English, *ib.*; the American, 252, *et seq.*; Bryant's, 252; Stoddard's and Whitman's, 252, 253; of some other poets, 253, 254.
"Eliot, George" (Mrs. Lewes-Cross), 137.
Elizabethan Period, songs from the dramatists, 170; poets of, their truth of life and character, 191; its imagination, 249; and see 75, 100, 105, 227; also

The Drama and *Dramatic Poetry.*
Elliott, Ebenezer, quoted, 126.
"Eloisa to Abelard," Pope, 214.
Eloquence, sometimes injurious to poetry, 59.
Emerson, on inspiration and insight, 23, 24; anecdotes of, 130, 153; his words and phrases, 242; on beauty and joy, 267; his "Threnody," *ib.*; on beauty, 149, 150; quoted, 130, 220, 221, 296; and see 35, 39, 50, 58, 75, 134, 136, 203, 213, 290.
Emotion, Wordsworth on, 20; Watts on, 26; the poet must be impassioned, 49; instinctively forms expression, *ib.*; its suggestion by music, 66; present call for, in art, 211; "uttered," 262, 263; and see *Passion.*
Empedocles, 212.
Empiricism, its service to the modern poet, 32.
Encyclopædia Britannica, article on Poetry, Watts, 26.
Endurance, the test of art, 166 *et seq.*; natural selection, 166, 167; of classic masterpieces, 168; of certain English poems, 170–172; "Ars Victrix," 173; transient aspects to be avoided, 201; of Shakespeare, 230, 231.
Endymion, Keats, quotation from its Preface, 122.
English Language, King James's Version of the Bible, 85; and see *Diction.*
English Poetry, and the imagination, 249, 250.

English Poets, Ward's, 249.
Environment, effects of, in youth, 9, 10; truth to, 199-201; of the Antique, 199; a lesson from Lowell, 200; home fields for art, *ib.*; transient conditions inessential, 201; one defect of Taine's theory, 276.
Epic Poetry, as evolved from folk-songs, 94, 95; the Homeric epos, 95-97; less inclusive than dramatic, 106; Firdusi's *Shah Nameh*, 111; the *Divine Comedy* of Dante, 112-115; Milton's *Paradise Lost*, 115-117; Arnold's epical studies, 133-135; Walter Scott, 135; simplicity of, 194; a growth, 237; and see *Objectivity*.
Epicureanism, 217.
Epicurus, 212.
Epigram, Latin, 92.
Equanimity, modern, 274.
Esther, The Book of, 175.
Ethics, of Homer, 95; truth of ethical insight, 216-219; the highest wisdom, 216; a prosaic moral repulsive and unethical, 216, 217; affected conviction, 216; why baseness is fatal to art, *ib.*; all great poetry ethical, 217,—and this whether iconoclastic or constructive, *ib.*; Shelley and his mission, 218, 246; and see *Truth*.
Euripides, his modern note, 88; and the Greek drama, 99; and see 137.
Evanescence, the note of, 181-185.
Eve of St. Agnes, The, Keats, 177.

Evolution, 287; and see *Science*.
Exaltation, national, 83; dramatic, 271.
Excursion, The, Wordsworth, 206.
Execution of the true artist, 235.
Executive Force, guided by the imagination, 228, 229.
Expression, chief function of all the fine arts, 44; as the source of beauty, 152; need of a free vehicle, 214; moved by imagination, 257; its poetic factors, 259; perfected by emotion, 261; should be inevitable, 274.
Ezekiel, quoted, 287.

FACILITY, undue, 235.
"Faculty Divine, The," so called by Wordsworth, 259; what it includes, 277.
Fairfield, F. G., neurotic theory of genius, 284.
Faith, the scientist's grounded in knowledge only, 33; and science, Lowell on, 37; Judaic anthropomorphism, 83; its indispensability, 280-296; recent lack of, *ib.*; distrust and cynicism, 289; works for distinction, 289-291; its poetic masterpiece, the Church Liturgy, 291-294; unrest of Arnold and Clough, 294, 295; the new day, 295.
Fame, Palgrave on popular judgment, 136; the case of Burns, 265; of Byron, *ib.*
Fancy, "The Culprit Fay," 236; the realm of, 247, 248; and see 215, 254.

Fantasy, distinguished from imagination, 236.
"Farewell to Nancy," Burns, quoted, 265.
Fashion, effect of, 39; the temporary must be distinguished from the lasting, 166; poetic style of Queen Anne's time, 213, 214.
Faust, Goethe, Snider on, 95; subjectivity of, 137; and see 104, 119, 137, 238, 269.
Faustus, Marlowe, 238.
Feeling, classical expressions of, 88 *et seq.*; of Wordsworth, 263, 264; its quality illustrated, 264, 265; religious, national, etc., 266; more accurate than thought, 286; and see 147; also *Passion*.
"Feigned History," as a generic term for all imaginative literature, 56, 57.
Felicia, Fanny Murfree, 208, 209.
Female Poets, Sappho, 88; Mrs. Browning, 88, 128, 266; subjectivity of, 127; Miss Lazarus, 266; Miss Rossetti, 269.
Femininity, self-expression the minor key of song, 127.
Ferishtah's Fancies, Browning, 225.
Fiction, prose (including *The Novel, Prose Romance*, etc.), letter from A. S. Hardy, 36; distinguished from the poetry under consideration, 56-59; as "Feigned History," 56; invention as to plot, narrative, characterization, 57; must not be rhythmical, 57-59; inborn gift of the great novelists and romancers, 60; Howells', *ib.*; as the principal outcome of recent dramatic and creative faculties, 137, 138; great modern novels and novelists, 137; the chief literary distinction of the century, 138; functions of the novelist, 237; examples of reserved power, 273.
Fielding, 60.
Fine Arts. See *Arts, The Fine*.
Finish. See *Technique*.
Firdusi, 111.
"Fitness of Things, The," 45, 156.
Fitzgerald, E., 82, 217.
Flavor, natural, 180.
Fletcher, J., 171.
"Flood of Years, The," Bryant, 252.
Folk Songs, 176; of Ireland and Scotland, 180; and see *Ballads*.
Force, that of poetry, 3,—not exerted by mere intellect and culture, 60; ethical, 217; as the vital spark, 259; and see INTRODUCTION.
Ford, John, 108.
"Forgiveness, A," Browning, 109.
Form, greatness of the dramatic, 107; English measures, 215; and see *Construction* and *Technique*.
Forman, H. Buxton, a phrase of, 52.
Formlessness of outline, its poetic effect. 246.
Foscolo, 133.
Fourier, C., 9.
Freedom of the poet's field, 220.

French Poetry, 18.
French Revolution, effects of, 123, 215.
French School, influence of the recent, 226.
Fuller, George, painter, 246.
Fuller, Margaret, anecdote of, 50.
Future, promise of the, 296.

GALILEO, 33.
Galland, Orientalist, 82.
"Gaspar Becerra," Longfellow, 220.
Gautier, Th., paraphrased by Dobson, 173; and see 120, 130, 179, 290.
Gebir, Landor, 205.
Genius, finds its natural sustenance, 10; vindication of, 46,— by Plato, *ib.*; scientifically defined by Hartmann, *ib.*; "poeta nascitur," 53; "a born lawyer," etc., 56; its methods imitable through industry and culture, 59–61; the dramatic, 104; the real test of poetry, 139; self training of, 145; its individuality the sole value of art, according to Véron, 152, 153; natural bent of, is unchangeable, 220; has a law of its own, 247; the question of its existence and nature, 276–285; recognition of, 277; as an inherent gift, *ib.*; Carlyle, Lowell, and Howells on, 278; distinguished from talent, 279,—from taste, 280; congenital, 281, 282; is original, 283; not dependent on theme, *ib.*; often limited; *ib.*; the universal type, health of, 284; its good sense, 284; should be obeyed by its possessor, 284, 285; its spontaneity, 285; inspiration of, 287; and see 147.
Georgian Period, 55, 116, 125, 138, 225, 227, 250.
Gérôme, painter, 239.
Gerusalemme Liberata, Tasso, 112.
Gesta Romanorum, 131.
Gilder, R. W., quoted, 257.
Gildersleeve, Basil L., 100.
Gnomic Poetry, 75.
Goethe, "Wertherism," 121; Heine's criticism of him and of Schiller, 18; and the Romantic movement, 119; Arnold's study of, 133; on art, 201; on epic and dramatic poets, 237; quoted, 64, 142, 143, 247; and see 54, 58, 76, 113, 118, 168, 263, 269, 290; and see *Faust*.
Golden Ages, 78.
Golden Treasury, The, Palgrave, 136, 172.
Goldsmith, 172, 250, 268.
Gosse, quoted, 111.
Gothic Art and Song, 159.
Grace, 179; and see *Charm*.
Gray, 172, 250.
Greek Bucolic Poets, 169.
Greek Dramatists. See *Dramatists, the Greek*.
Greek Lyrists, 169.
Greeks, The. See *The Antique, Objectivity*, etc.
Gregory VII., Horne, 104.
Grotesque, The, 160.
Guy Mannering, Scott, 137.

"HAMADRYAD, THE," Landor, 200.
Hamlet, Shakespeare, 102, 104.
Harris, W. T., inspirational view of poetry, 23; and see 174.
Hartlib, S., Milton's friend, 27.
Hartmann, E. von, metaphysician, on genius, 46, 282; on the idea in art, 156.
Hawthorne, 137, 218, 273.
Haydon, B. R., painter, on art, 201.
Hazlitt, W., logical view of poetry, 25; cited, 207, 208.
Health, recoverable in poetry, 295.
Hebraism, 99, 290; and see *Bible, Poetry of the*.
Heine, compared with Byron, 125; character and genius, 126; his mocking note, 127; quoted, 112; cited, 140; and see 135, 142, 203, 208.
Hellenism, Landor, 124; compared with Latinism, 90, 91; effect on Vergil, 91; Poetry of Greece, 87-90, 95-100; the Greek lyrists, 87; the anthology, etc., 88, 89; idyllists, 89, 90; the Homeric epos, 95-97; the Attic dramatists, 97-100; antique view of tragedy, etc., 103, 104.
Henry Esmond, Thackeray, 55, 137.
Heredity of genius, 277.
Herodotus, 169.
Heroic poetry, Horace's view of, 18.
Herrick, 168, 171.
Hesiod, and Vergil's Georgics, 91.
"Highland Mary," Burns, 265.
Hindu Literature, 81, 82.

Hoffmann, Ernst, 142.
Holinshed, chronicler, 57.
Holmes, quoted, 281.
Homer, Lord Derby's version, 82; Vergil's obligations to, 91; Snider's ethical theory of the Iliad and Odyssey, 95; his joyous and perfect transcript of life, 96, 111; highest value of, 97; modern "Homeric Echoes," 134; Arnold on the swift epic movement, 134; descriptive touches of, 190; ethics, 217; endurance, 230; impassioned characters, 270, 271; quoted, 194; and see 78, 106, 106, 191, 236, 251, 269, 290.
Hood, a poet of emotion, 265.
Horace, concerning poetry, 17; progenitor of the *beaux esprits*, 93; and see 27, 169.
"Horatian Ode, An," Stoddard, 239.
Horatii, The, 93.
Horne, R. H., *Gregory VII.*, 104; dramas of, 132; and see 29, 133.
Household Book of Poetry, The, Dana, 236.
House of Life, The, Rossetti, 269.
Howe, Julia Ward, 267.
Howell, Elizabeth Lloyd, 266.
Howells, W. D., as illustrating both natural gift and training, 60; on recent Italian poetry, 128, 129; on genius as "natural aptitude," 278.
Hugo, V., *Hernani*, 104; and the Romantic movement, 119; and see 18, 133, 142, 269, 287, 290.

Human Element, The, 269; of the Liturgy, 292 *et seq.*
Humor, as a pathetic factor, 215; and see 123, 275.
Hunt, Leigh, on poetry, 25; cited, 239; quoted, 243; and see 119, 173, 179, 225.
Hutchinson, Ellen M., quoted, 182.
"Hymn, before Sunrise," Coleridge, 266.
"Hymn to Aphrodite," Sappho, 88.
Hyperion, Keats, its large utterance, 248; quoted, *ib.*

IBSEN, 42.
"Ichabod," Whittier, 268.
Idealist, The, how affected by the new learning, 34–37.
Ideality, its bearing on individual action, 3–5; present struggle with empiricism, 34–39; the "shews of things" are real to the poet, 153; characterizes true Realism, 199; opposed alike to prosaic goodness and to vice, 216; against impurity, 262; present want of, 274.
Ideals, racial and national, 159, 162–165; the Aryan, 159; Academic, 161; the Japanese, etc., 162–165.
Ideal, the artist's, what constitutes it, 41.
Idyllic Poetry, Keats, Tennyson, etc., 68, 69; of the Bible, 87; *Ruth* and *Esther* contrasted with *Anna Karénina*, 175; Tennyson's method, 193; recent idyllic period, 210, 211; *Snow Bound*, 268; over supply of, 275.
Idyllic quality, 225.
"I have loved flowers that fade," Bridges, 185.
Iliad, The. See *Homer*.
"Il Penseroso," Milton, 116.
Imagery, when outworn, 34; of poets, Joubert on, 235.
Imagination, sovereign of the arts, 5; its office fully recognized by Wordsworth and Coleridge, 20, 21; Schopenhauer on, 21; indispensable to the savant, 32; the savant's akin to the poet's, 36; increased material for, 38; the dramatic, 104; nothing forbidden to it, 201; glorifies Shakespeare's errors of fact, *ib.*; of the intellect, 211; freed by a free rhythm, 214; *considered as the informing element of poetic expression*, 225–258; lack of, in recent poetry, 227; chief factor in human action, 228; the executive, 229; the poetic, *ib.*; Shakespeare's, 229–231; *definition of*, 231, — illustrations of same, 232–235; how to test it, 232; must be clear, *ib.*; must have "holding power," 233; of Blake, *ib.*; definiteness of, 234; not confined to the superhuman, 236; its higher flights, 236, 237; when inventive and constructive, 237; when purely creative, 237, 238; its Wonderland, 238; its power of suggestiveness and prevision, 239; im-

aginative diction, 240–242; of the supernatural, 243, 244; sublimity of the Vague, 244–247; of Shelley, poet of cloudland, 246; effect of magnitude on the, 247; how distinguished from Fancy, 247; "the grand manner," 248; of the Elizabethans, 249; usually deficient between the Elizabethan and Georgian periods, 250; comparison, association, etc., 250; of "elemental" bards, 250–254; divinity of this creative gift, 254–258; as used and excited by emotion, 261; imagined feeling, 273; promised revival of, 296; and see 147, 166.

Imitation, normal in youth, 13; how far the office of art, 17; Vergil's, of the Greek poets, 91; Longfellow's, 91, 92; Schlegel on, 92; the greatest work inimitable, 109; servile, not true realism, 197, 198.

Immaturity, posing of Heine's youthful imitators, 127; and see *Training*.

Immorality. See *Ethics*.

"Impassioned," its meaning, 261.

Impersonality. See *Objectivity*.

Impressionism, of the Nineteenth Century poets, 118; true, 144; how allied to transcendentalism, 149; its value and defects, 150; and see 153.

Individuality, of certain writers, 58; of style, 80; Longfellow's specific tone, 92; Browning's, 108; its loss means death in art, 144; national, how lost, 164; the poet's distinctive voice, 297; and see *Subjectivity*.

Industry, differentiated from faculty, 46; of men of genius, 278.

In Memoriam, Tennyson's, 55, 212, 225, 270.

Inness, G., painter, 246.

Insight, the poet as a seer, 22–24; preceding demonstration, 37; the source of revelation, 45, — Plato and Wordsworth on, *ib.*; allied with genius, 46; the celestial, 65; the Miltonic, 116, 117; nature of, 285; as inspiration, 287; and see 147.

Inspiration, as a poetic factor, according to Plato and his successors, 21–24, 46; Zoroaster on, 22; Hebraic, 75; pseudo, 235; belief in direct, 287; the prophetic, 287; and see 147.

Instinct, 284.

Intellectuality, Browning's, 109; Milton's learning, 116; poetry of wisdom and morals, 211–215; and see *Thought*.

Intelligibility, 236.

Intensity, of emotion, 261; the dramatic, 273.

Interpretation, of nature, see Section VI., *passim*; Wordsworth, Bryant and the American School, 195; Whitman's and Lanier's, 195, 196; subjective, of Nineteenth Century poets, 202–204; the pathetic fallacy, 204–210; and see *Revelation, Insight*, etc.

Interpretative Faculty, 26; and see *Insight*, etc.
Inter-Transmutation of certain poetic styles, 215.
INTRODUCTION. Account of the origin, purpose and method of the present treatise, pp. vii–xvii.
Introspection, conventual, 140; and see *Subjectivity*.
Intuition, unconscious process of the soul, 147; superior to logical process, 157; should be obeyed, 284; woman's, 286.
Invention, fiction its modern outlet, 137, 138; an imaginative function, 237.
Ion, Euripides, 99.
Irony. See *Satire*.
Irreverence, dangers of artistic, 158.
Isabella, Keats, 239.
Isaiah, 236.
Island, The, Byron, 206.
"Israfel," Poe, 73.
Italian influence, 162.
Italian poetry, English obligations to, 115; of modern Italy, 128.

JAMES, G. P. R., novelist, 137.
James, H., novelist, 192.
Japanese, the, artistic method of, 31; their literature, 81; antipodal art ideals of, 162–165; they recognize fitness and ideal beauty, 162, 163; danger menacing their individuality, 164; assimilative tendency of, 165.
"Jeanie Morrison," Motherwell, 265.
Job, The Book of, its grandeur and impersonality, 86; quoted, 244; and see 36, 104, 113, 236.
Johns Hopkins University, its origin and founder, 3, 4; and see 61, 93.
Johnson, Dr., 14.
Jones, Sir William, 82.
Jonson, Ben, on language, 51; on the English stage, 110; plays and songs of, 170; and see 250.
Joubert, critic, 15, 135; quoted, 143.
Joy of the poet, 267.
Judaism. See *Bible, Poetry of the*.
Judgment, 284.
Julius Cæsar, Shakespeare, 104.

KEATS, quoted, 67; as an artist-poet, 68, 69; dramatic promise of, 69, 110; his place in English poetry, 110; on immaturity, 122; creative works of, 124; on beauty and truth, 187, 220; imaginative diction of, 241; his style, 248; on intensity, 262; quoted, 182; and see 10, 60, 75, 116, 119, 133, 172, 173, 177, 193, 225, 239, 255.
Kepler, astronomer, quoted, 155.
Kingsley, *Hypatia*, 118.
"Kubla Khan," Coleridge, 125.

LA FARGE, J., artist, cited, 78, 79; on Japanese Architecture, 163.
Lake School, The, 33.
"L' Allegro," Milton, 116.
Lamartine, 120.
Lamb, C., 162, 170.
"Land o' the Leal, The," Lady Nairne, 265.

Landor, creative works, 124; his "sea-shell," 205, 206; cited, 18; quoted, 184, 204, 239, 241, 247; and see 58, 82, 133, 135, 179, 200, 203.

Landscape. See *Descriptive Poetry*, *Nature*, etc.

Lang, A., renderings from the Anthology, 89; and see 82.

Language, its efficacy to express ideas, 15, 16; poetry absolutely dependent on, for its concrete existence, 50; Ben Jonson on, as speech, 51; must become rhythmic to be minstrelsy, 51-55; "idealized language," 52; vibratory power of words, 52; a test of genuineness, 54; speech a more complex music than music itself, 72, 179; the Hebrew, 87; the genius of our English tongue, 214, — its eclecticism and increase, 215; and see *Diction*.

Lanier, compared with Whitman, 196; his imagination, 253; musical genius of, 282; his work tentative, *ib.*; and see 62, 158.

Laocoön, Lessing, 66.

Latinism, sentiment of Latin poets, 90-94; Vergil and his modern countertypes, 91; Ovid, Catullus, etc., 92; *Tu Marcellus eris*, 93; Horace and the Horatii, 93, 94.

Law, natural, the working basis of all art, 6-8; poetic, 62.

Lay of the Last Minstrel, The, Scott, 131, 238.

Lazarus, Emma, 266.

Learning, the New, 34; and see *Science*.

Lee-Hamilton, E., sonnet by, 206.

Leighton, Sir F., painter, 279.

Leopardi, 133.

Les Précieuses, Molière, 100.

Les Trois Mousquetaires, Dumas, 137.

Life, conduct of, 5; the poet supreme among artists in the portrayal of, 70, 71.

Life School, prospective rise of a, 211.

Light of Asia, The, E. Arnold, 82, 235.

Limitations, of specific genius, 80, 283; charmingly observed by the Horatii, 94; and see 288.

Lincoln, Abraham, 143.

"Lines to an Indian Air," Shelley, 266.

Liszt, musician, 9, 232.

Literary eras. See *Periods, literary and artistic*.

Liturgy, The Church, as a literary masterpiece of Faith, 291-294; its universal and human quality, 292; symphonic perfection, 293; its uniqueness, *ib.*

"Local" Flavor, Lowell's taste for American lyrics, 200; a home-field for our sculptors, *ib.*

"Locksley Hall," Tennyson, 270.

Lodge, O. J., physicist, 35.

Lombroso, C., a theory of, 284.

Longfellow, the New World counterpart of Vergil, 91; "Gaspar Becerra," 220; his national po-

etry, 268 ; quoted, 30, 260 ; and see 136, 203, 225, 235, 289.
Longinus, cited by Dryden, 18.
Long Poem, A, is the designation a misnomer? 178.
Lorna Doone, Blackmore, 137.
"Lost Occasion, The," Whittier, 268.
"Lotos-Eaters, The," Tennyson, 177.
Love as a master-passion, 260.
Lovelace, R., 167, 171.
Lowell, on faith and science, 57 ; on Addison and Steele, 100; his national sentiment, 129 ; his truth to nature, 190 ; his respect for "local" flavor, 200; on our view of nature, quoted, 207 ; verbal aptness of, 242 ; Odes, 267; on talent and genius, 278; quoted, 144, 180, 188 ; and see 4, 27, 136, 162, 195, 203, 213.
Lucile, Lytton, 237.
Lucretius, 75, 91, 212, 217.
"Lucretius," Tennyson, 270.
Lusiad, The, Camoëns, 112, 244.
Lyall, Sir A., quoted, 182.
"Lycidas," Milton, 90, 116 ; quoted, 102.
Lyrical Ballads, Wordsworth and Coleridge, 21.
Lyrical Poetry, not always subjective, 83 ; the Davidic lyre, 84, 85 ; early alliance with music, 85 ; the Greek lyrists, 87, 88 ; Catullus, *et al.*, 92 ; Horace and his successors, 93 ; primitive ballads, etc., 94 ; songs of the drama, 105 ; Ariel's songs,

107 ; characteristics of the pure lyric, 178-180 ; and see 264, 265, also *Songs* and *Lyrics*.
Lytton, Robert, Lord, *Lucille*, 237.

MACAULAY, on poets as critics, 12
Macpherson, 58.
Mahaffy, J. P., scholar, 88.
Manzoni, 210.
Marlowe, 167, 238, 249.
Marmion, Scott, 131, 135.
Marseillaise, La, De l'Isle, 266.
Martin, Homer, painter, 246.
Martineau, Harriet, quoted, 76.
Mary Stuart, Swinburne, 132.
Masculinity, impersonality the major key of song, 127.
Masque of the Gods, The, Taylor, 254.
Masque, The (Elizabethan), 107.
Masterpieces, appeal to the public and the critical few, 197 ; the Church Liturgy, 291 *et seq.*
Masters, the, their influence on youth, 10.
Materialism, 3 ; Whitman on, 38.
Materials, poetic, not a substitute for imagination, 235.
Maxwell, C., scientist, 35.
Mediocrity of followers in art, 151.
"Meditations of a Hindu Prince," Lyall, 182.
Meleager, 89.
"Melencolia," Dürer, 140 (and see Frontispiece).
Melodramatic quality, Hugo, 119.
Melody, as heard or symbolized, 174; of speech and of music, 179 ; the "dying fall," 183.
Men and Women, Browning, 109.

Menzel, critic, cited, 118.
Meredith, G., 42, 137.
Metaphysics, hostile to a true æsthetic, 149; effect upon Coleridge, 239; and see *Didacticism*.
Michelangelo, 50, 101, 233, 247, 272.
"Midsummer Meditation, A," Gilder, 257.
Midsummer Night's Dream, Shakespeare, 189, 201, 248.
Mill, J. S., on philosophy, 16; on poetry as emotion, 19, 20; on feeling and rhythm, 55; on the novelist and the poet, 138; on feeling and thought, 261; and see 178.
Millais, Sir J., painter, 279.
Miller, "Olive Thorne" [Harriet M.], 62.
Millet, J. F., painter, 10, 255.
Milton, his poetic canon, 27, 49, 260; tractate *On Education*, 27; as a Copernican, 35; his prose, 59; early poems of, 168; *Comus*, 236; his Satan, 238; his imaginative words and phrases, 240 *et seq.*; his imagination, 245; quoted, 53; and see 58, 76, 90, 113, 172, 193, 226, 248, 250, 269, 287, 290.
"Miltonic," 240.
Miltonic Canon, the, its constituent of passion, 260, 261; and see 27, 49.
"Milton in his Blindness," Howell, 266.
"Mimnermus in Church," Cory, 183.

Minor Poets, 80; the creative masters in youth, 101; recent, their merits and defects, 136, 226, 227, 254; the new art-school, 173.
Mirth, 275.
Mixed type, poetry of the, 113, 114.
Molière, 100, 283.
Moore, T., states the Byronic creed, 19.
Monet, Claude, painter, 158.
Monodramas, 109.
Monticelli, painter, 158.
Moralism. See *Didacticism*, etc.
More, Henry, quoted, 286.
Morris, W., contrasted with Walter Scott, 131; and Chaucer, 170.
Moschus, paraphrase on a passage in Job, 90; Epitaph on Bion, 90.
Motherwell, W., 266.
Motive, recent lack of, 289.
Mozart, musical genius of, 281.
Murfree, Fanny N. D., novelist, quoted, 208, 209.
Music, as a sensation, 15; Lanier's devotion to, 62, 282; its range and limits of expression, and as compared with poetry, 64–66; special functions, 64; is it the highest art? *ib.*; Poe on, 65; Schopenhauer and Spencer on, 65; expresses feeling, not thought, 66; its effect on the rhythm, etc., of the Hebrew psalms, 85; philosophy of, 179; the musician's memory, 232; and see 264.

ANALYTICAL INDEX 321

Musset; 120, 290.
Muybridge, E. J., his instantaneous photography, 198, 199.
Myers, F. W. H., on genius, 282.
"My Last Duchess," Browning, 109.
"My Maryland," Randall, 266.

NAIRNE, LADY, 264.
Naïveté, of the Psalms, etc., 85; and see 180, also *Naturalness*.
Narrative poetry, inferior in realism to dramatic, 107; see *Objectivity*.
National sentiment, the modern Italian, 128; of American verse, 129.
Naturalism, 197; the true, 273.
Naturalness, excellence of genuine feeling, 142; return of, 173; makes for simplicity, 176; affectation of, 177; method should seem unconscious, 193; and see 264.
Nature, trains the poet, 10; Milton's treatment of, 116; normal beauty of, 156, 157; poetry of, from Wordsworth and Bryant to Lanier, 194-196; subjective in recent times, 202-204; the modern return to, 204; does she give solace and sympathy? 204-209; full of motion and unrest, 208; the sovereign of modern art and song, 210; her triumph too prolonged, 211; universally set forth by Shakespeare, 229; and Wordsworth's similes, 250.

Neo-impressionism, 153; and see *Impressionism*.
Neo-Romanticism, 130.
"Neurotic disorder," the question of, 284.
Newcomes, The, Thackeray, 137.
Nibelungen Lied, 131.
Nineteenth Century, literary eminence of, 138; its idealization of Nature, in art and poetry, 210; Wordsworth's place in, 219.
Norse poetry, sages, 78.
Notre Dame de Paris, Hugo, 137.
Novalis, 142.
Novel, the, and Novelists. See *Prose Fiction*.
Novels in verse, 237.
Novelty, romantic effect of strangeness, 151; stimulates zest, 160.

OBJECTIVITY, creative and impersonal poetry, 75-110, *passim*; absolute vision, 77, 78; creative eras, 79; of the Book of Job, 86; the Hebrew idyls, 87, 175; primitive ballads, 94; charm of, in the antique, 96, 97; the Greek drama, 97-101; the dramatic genius, 104; impersonality of the old masters, 107; Homer, 111; Chaucer, 115; Burns, 120; its restorative charm, 121; of certain productions of Shelley, Keats, Landor, 124; masculine, and in the major key, 127; pseudo-impersonality of artistic recent verse, 130, 131; Walter Scott, 131;

Arnold's theory and epical studies, 133, 134; the "note of the inevitable," 134; our greater prose fiction, 137, 138; not the chief test of poetic genius, 139; of the antique, compared with the muse of Christendom, 139-143; its invigorating value, 142; not the final test of poetry, 144; may be tame and artificial, *ib.*; the antique view of nature, 207; creation of impassioned types, 270-273; and see 250.
Obscurity, 235.
"Ode on a Grecian Urn," Keats, 67, 187.
Odes, Lowell's and Emerson's, 267.
Odyssey, The, 131; and see *Homer*.
Œdipus at Colonos, Sophocles, 190, 238.
Œdipus Tyrannus, Sophocles, 98, 104.
"Œnone," Tennyson, 177, 200, 242.
"Old Pictures in Florence," Browning, 170.
Omar Khayyám, Rubáiyát of, Fitzgerald, 82, 212.
"On a Bust of Dante," Parsons, 254.
On Education, Milton's tractate, 27.
On the Heights, Auerbach, 137.
"On the Receipt of my Mother's Picture," Cowper, 274.
Optimism, of sovereign poets, 290; and see 295, 296.
Orientalism, Zoroaster, 22; Japanese art, 31; the Asiatic inspiration, how far understood and transformed by us, 81, 82; Hebrew genius and poetry, 82, 87; influence on the Alexandrian school, 90; Firdusi, 111; India, China, etc., 162; and see 244, also *Hebraism, the Japanese, Poetry of the Bible*, etc.
Originality, distinguished from skill, 60; not discordant with universal principles, 151; of genius, 277, 283.
Orion, Horne, 29.
Orlando Innamorato, Ariosto, 112.
Outline, 246.
Ovid, cited, 19; and see 92.

PAGANISM, 112; and see *The Antique*.
Painting. Guido's Aurora, 29; the Oriental and the Western methods of vision, 31, 32; the born painter, 53; its powers and limits, 63; must avoid a literary cast, 71; superior to poetry in depicting visible nature, 202; and see 282.
"Palace of Art, The," Tennyson, 167.
Palgrave, F. T., quoted, 136; and see 172.
Paradise Lost, Milton, 35, 238, 245.
Paradiso, Il, Dante, 174.
Parnassiens, the French, 130.
Parsons, T. W., "On a Bust of Dante," 114; and see 254.
Passion, The Romantic view of

Poetry as the lyrical expression of Emotion, 19, 20, 262; exalted national feeling, 83; intensity of Hebraic emotion, 83-85; Sappho, 87, 88; of Mrs. Browning's sonnets, 128; "Eloisa to Abelard," 214; subdued in modern poetry, 227; as the force and excitant of imaginative expression, 257, 260-276; required by the Miltonic canon, 27, 260, 261; defined, *ib.*; not love alone, 260; as intense emotion, 261; its use of imagination, *ib.*; must be genuine and pure, 262; modern understanding of, 262, 263; of Wordsworth, its limits, 263; as Feeling and Sentiment, 264, 265; of women poets, 266; of ardor, joy, grief, etc., 266-268; Whittier's, 268; as art's highest theme, *ib.*; its human element, 269; Tennyson's, in youth and age, *ib.*; creation of its objective types, 270-272; exaltation through intense sensations, 271, 272; reserved power of, 273; absolute dramatic, 273, 274; its occasional lulls, 275; excited by great occasions, 276, 288; of the cries of Faith, 292; and see 5, 166, also *Emotion, Feeling*, and *The Romantic School*.
Pastoral Verse. See *Greek Bucolic Poetry, Nature, Idyllic Poetry*, etc.
Pater, W., quoted, 159.
"Pathetic Fallacy," the, Ruskin's phrase and explanation, 204; consideration of, 204-210; whether our feeling concerning nature is an illusion, 205; illustrated by Landor and Wordsworth's treatment of the sea-shell's murmur, 205, 206; Lee-Hamilton's sonnet on, 206; Lowell on, 207; the "illusion" likely to be cherished, 209.
Patrician Verse, "The Rape of the Lock," 214.
Pattison, Mark, on poetical prose and the prose of poets, 58; quoted, 268.
Payne, John, 82.
Pentameron, The, Landor, 125.
Percival, J. C., 190.
Père Goriot, Balzac, 137.
Pericles and Aspasia, Landor, 124, 125.
Periods, Literary and Artistic, how to determine their quality, 226; reactionary, 275; the older American, 276; heroic, culminating, etc., *ib.*; and see *Alexandrian Period, Composite Period, Elizabethan Period, Georgian Period, Queen Anne's Time, Victorian Period*, etc.
Persia, Poetry of, 111.
Personality. See *Subjectivity*.
Pessimism, 291.
Pheidias, 150.
Philip Van Artevelde, Taylor, 104.
Philistinism, British, 123; Heine's revolt against, 126; and see 222, 290.
Philosophical Poetry,—that of wisdom and ethics. See *Didacticism, Ethics*, and *Truth*.

Photographic Method, lessons taught by Muybridge's camera, 198, 199; not to be closely followed, 199; and see *Realism*.
"Pied Piper, The," Browning, 215.
Pilot, The, Cooper, 137.
Pindar, not strictly subjective, 83, 87; and see 87, 142, 251.
Pippa Passes, Browning, 55.
Plato, and Aristotle, 17; his conception of poetry and the poet, 21-24; "The Republic," 21; a poet-philosopher, 22; his pupils, 22-24; on insight, 45; on inspiration, 46; and modern transcendentalism, 149; and see 20, 57.
Platonism, its drawbacks, 24; and see *Plato*.
Plautus, 100.
Plot, 174.
Plotinus, 23.
Poe, on the form of words, 15; definition of poetry, 26, 151 *et seq.*; on music, 65; quoted, 73; his passion, 267; and see 133, 178, 183, 242, 283.
Poet, the, his freedom, 20; Platonic idea of, 21; Plato's banishment of, from The Republic, 21, 22; his two functions, 28 *et seq.*; how affected by the new learning, 34-36; compared with the savant, 36, 37; his province inalienable, 38; a creator, 44; a revealer, 45; his power of expression, 47; his wisdom, 48; sensitiveness, 49; must be a born rhythmist, 53; as a writer of prose, 57, 58; pseudo-poets, 60; must be articulate, 62; Lessing on, 71; may use all artistic effects, *ib.*; Emerson, 149; a phenomenalist, 155, 156; sees and restores the beautiful, 157; expresses his true nature and his work, 167; his youthful passion for the beautiful, 168; his rendering of nature, 188 *et seq.*; cannot depict landscape like the painter, 202; nature's subjective interpreter, 202, 203; the coming poet, 211; Pope 213, 214; Shelley, 218; his final recognition of beauteous verity, 220, 221; his religious point of view, 221-223; an anecdote, 234; Joubert on the true poet, 235; rarely a sensualist, 246; his imaginative realism, 254; his godlike creative gift, 256, 257; his vital spark, 259; Mill on, 261; Poe and Emerson, 267; "of the first order," 268; his line of advance, 269; Stendhal on, 272; how begotten, 276; elder American poets, *ib.*; must have the "faculty divine," 277; genius of, 277-285; Carlyle on, 278; originality of, 283; his noble discontent, 286; the Vates, 287; his modern distrust, 288 *et seq.*; his station higher than the critic's, 297; and see *Female Poets, Poetry*, Ποιητής, etc.
Poetic Principle, The, Poe's lecture, 26.
POETRY, as a force in human life

and action, 3; why sometimes esteemed too lightly, 5, 14; to be observed scientifically, 6, — and in the concrete, 8; why its definition is needed, 8; it is vocal, 9; as a sensation, 15; *historic views and definitions of*, 17-28; the antique view,—Aristotle, Horace, 17; Chapman, Dryden, Landor, Goethe, 18, — Arnold, 19, — traversed by Heine, 18; the Romantic, or Emotional view, 19, 20, — Byron, Moore, Mills, 19, — Bascom, Ruskin, 20; not opposed to Prose, *ib.*; stress laid upon Imagination by Wordsworth and Coleridge, 20, — by Schopenhauer, 21; the Platonic view, — Plato in "The Republic," etc., 21 *et seq.*; Zoroaster cited, 22, — Cicero, 22, — Bacon, Sidney, Plotinus, Carlyle, Emerson, Harris, 23, 24; partial failure of all statements, 24, 25; clearer modern views, — the artistic, Hazlitt, Hunt, Shelley, Watts, 25, 26, 28; the æsthetic view, — Poe, 26; a phrase of Milton, 27, — of Arnold, 27; the statement still incomplete, 28; poetry as the antithesis to Science, 28, — what this means, *ib.*; illustrated, 29, 30; effect of exact science on, 33, 37; Professor Hardy's view of, 36; Tyndall on, 39; *Defined and examined in relation to the other Fine Arts*, 41-73; the present a fit time for its consideration, 42; its *spirit* not reducible to terms, 42; a force, 43, which enters the concrete, *ib.*; DEFINITION OF POETRY IN THE CONCRETE, 44; a creation, through invention and expression, 44; a revelation, through insight, 45, 46; as an expression of beauty and truth, 46-48; as an expression of intellectual thought, 48, — of emotion, 49; as eminently an art of speech, 50; its language essentially rhythmical, 51-56, 62; its vibratory thrills, — their rationale, 52; rhythmical factors of, 54; its rhythmical impulse spontaneous, and equal to the degree of emotion, 54, 55; how different from other forms of creative expression, 55, — from imaginative prose fiction, 56, — from rhetoric, eloquence, etc., 59; rhyme, etc., 56; often, however, may be cast in rhythmical prose form, 58; conforms to law, consciously or otherwise, 62; must be articulate, 62, 63; compared with music and the arts of design, 63-71; closely allied with music, 64; its achievements and limitations with respect to sculpture, 67, — to painting, 68; surpasses the rival arts by command of vocal movement, thus infusing Life, 69-71; compact analysis and summary of, 71, 72; universal range of, 75; *divided into two main streams, — the impersonal, or creative, and the personal, or self-*

expressive, 76-81; its technical partition, into the epic, dramatic, lyric, etc., 76; impersonal or unconditioned song, 77-81, 94-101, 104; creative masterpieces, 78, 79; self-expressive or subjective song, 80 *et seq.*, and 111-145, *passim*; eastern Asiatic, 81; Hebraic, 82-87; Hellenic, 87-90; Latin, 90-94; must not be disillusionized, 96; the grand drama, 101-107; modern and subjective drama, 108-110; Persian, — Firdusi, etc., 111; Italian and Portuguese epics, 112; the "Divina Commedia" and Dante, 112-115; allegorical, — "the Fairie Queene," 114; from Chaucer to Milton, 115; the great Puritan epic, — "Paradise Lost," 115-117; poetry of the Nineteenth Century, 118; the Romantic Movement, 118-120; Goethe, Hugo, etc., 119; of Burns, 120; of Byron, 120-123; Wertherism, 121; of sentiment in youth, 122; of Shelley, Keats, Landor, Coleridge, etc., 123-125; of Unrest, — Heine, 125-127; its masculine and feminine elements, or the major and minor keys of lyric song, 127; of Mrs. Browning, 128; of national sentiment, 128; of art for art's sake, 129 *et seq.*; of the Pre-Raphaelites, Parnassiens, Neo-Romanticists, 129, 130; latter-day verse, 130, 131; of Scott and W. Morris, contrasted, 131; of Swinburne, 131-133; of M. Arnold, in view of his early theory, 133-135; of the composite era, 136; of the antique, and that of Christendom, — an estimate of our loss and gain, 139-143; how it must be tested, 144; *as an artistic expression of the beautiful*, 147-185; Poe's definition of, 151, 152; movements for greater freedom and variety in, 158; translations of, 166; made enduring by beauty, through natural selection, 166-172; concrete beauty of, 167, 168; survival of classic masterpieces, 168, 169; beauty of our English, 170-172; modern art school of, 173; elements of its concrete perfection, 173-180; primitive rhapsody, 175; the *vox humana*, 178; Mill and Poe on "a long poem," 178; the pure lyric, 178-180; its note of evanescence, 181-185; the didactic heresy, 187; *element of Truth in*, 187-223; "description," its strength and weakness, 189, 190, 202; breadth of, *vs.* analysis, 191-193; naturalness, 193; "reflection," of nature, 194-196; realistic, 196-199; local flavor, 200; subjective expression of nature, 202-204; "the pathetic fallacy" in, illustrated by Landor, Wordsworth, etc., and refuted by Lee-Hamilton, 204-207; of nature, its modern importance, 210; a life-school needed, 211; of wisdom, the higher philosophical, 211-217; elements of humor in, 215; true, yet free, 219, 220; lack of imagination and passion in recent, 225-227; other modern traits of, 225, 226; *the imaginative element in*, 225-258; "spasmodic," 235; obscurity, *ib.*; its peopled wonderland, 238; suggestive, 239; diction of, 240 *et seq.*; supernaturalism in, 236, 243; of the Vague, 246; of Fancy, 247; English, 249; "elemental," 250-254; passion its incentive, 257; *as product of "the faculty divine,"* 259-297; *element of Passion in*, 260-276; Wordsworth's statement of, 263; of Scotland, 264; of English sentiment, 265,— of American, 267, 268; of intense emotions and impassioned types, 270 *et seq.*; true naturalism of, 273; the absolutely dramatic, 274; of heroic crises, 276, 288; *Genius*, 277-285; *Insight, Inspiration*, etc., 285-288; of Prophecy, 287; *Faith* indispensable to, 288-294; of the Church Liturgy, 291 *et seq.*; future of, 296; concerning the study of, 296, 297; its present dissemination, 297; and see INTRODUCTION, *passim*, also *Analytic Poetry, Bible, Christendom, Christianity, Descriptive Poetry, Dramatic Poetry, Elegiac Poetry, Epic Poetry, Gnomic Poetry, Heroic Poetry, Idyllic Poetry,* *Lyrical Poetry, Narrative Poetry, Norse Poetry, Orientalism, Reflective Poetry, Society-Verse, etc.*

POETS OF AMERICA, by the author of this volume: references to, 35, 101, 137, 160, 190, 211, 226, 246, 252, 268.

Ποιητής, Aristotle on, 17.

Pope, Byron on, 19; question of his genius, 213-215; his didacticism, 213; was he a poet? *ib.*; compared with modern leaders, 215; and see 116, 172.

Prayer Book, the Episcopal. See *The Church Liturgy*.

Pre-Raphaelitism, concurrent in various arts, 50; and see 158, 225.

Pretension of would-be genius, 280.

Prevision of the imagination, 239.

Primitive Poetry, sagas, folk-lore, ballads, etc., 78.

Prince Deukalion, Taylor, 254.

Princess, The, Tennyson, 237, 264.

Prior, 94.

"Problem, The," Emerson, 55.

Procter, B. W., 179.

Production, rather than motive, the test, 169.

Prometheus Bound, Æschylus, 98, 104.

Prometheus Unbound, Shelley, 124, 132.

Prophetic Faculty, the poet as Vates, 23; the Hebraic, 84; of Blake, 234; its vision, 285; and see 287, also *Inspiration*.

Prose, the antithesis of verse, 20;

injured by use of poetic rhythm, 57, 59; the prose of poets, 58, — of various rhapsodists, *ib.;* the rhythmical prose form, *ib.;* Arnold's, 294; and see *Fiction.*
Protagonists of the grand drama, 104.
Provençal Poetry, 168.
Prudhomme, Sully, 131.
Psalms, the Hebrew, 84, 85.
Psychical Quality, 177.
Psychological School, 80.
Public Opinion, ultimate, to be respected, 277.
Purpose, must be earnest, 288, 289.
Puttenham, George, quoted, 198.

QUALITY, the grace of lyric poets, 178.
Queen Anne's time, 213, 214.
Quintus, 169.
Quotations, miscellaneous, 30, 90, 140, 222, 285, 296.

RACE, 81; Latin traits distinguished from the Greek, 90; Latin quality and Gothic, 264; and see *Hebraism, Hellenism, Japanese,* etc.
Randall, J. R., 266.
"Randolph of Roanoke," Whittier, 268.
Rape of the Lock, The, Pope, 214, 215.
Raphael, 150; cited, 197.
Raphaelitism, Academic, 157.
Ratiocination. See *Analytic Poetry.*
Reaction. See *Periods, Literary.*
Realism, dramatic, 107; lacking in Milton's early poems, 116; and romanticism, 145, 199; and beauty, 150; descriptive details, 190; of the Elizabethan drama, 191; Whitman's, true and false, 196; not a display of facts, *ib.;* not a servile imitation, 197; Tennyson on, 198; must conform to human perceptions, 198, 199; photography, 199; should be idealized, *ib.*
"Reapers, The," Theocritus, 179.
Reason, or intellectual method, 284.
Recent Poetry, characteristics of, 288, 289.
Reflective Poetry, a modern type, 76; of Wordsworth, 189; of Nature, Wordsworth and Bryant, 194, 195; Wordsworth's self-contemplation, 203, 204; modern speculation, 207; and see *Truth, Didacticism,* etc.
Religion, its "conflict with science," 33; piety of the artist's labors, 221, — his conception of Deity, 222; the God of truth also the God of art, joy, song, 222, 223.
Renaissance, 112; tentative revolts and new movements in art, 158-161; the Italian, 160; the "Colonial" revival, 161.
Repose in art and poetry, 273.
Republic, The, Plato's, 21, 22.
Republicanism, its "applied" imagination, 229.
Reserve, professional, cause of, 12-14.
Reserved Power, 272, 273.

ANALYTICAL INDEX 329

Retentive Faculty. See *Imagination*.
Revelation, of scientific truth, 33 *et seq.*; through poetic insight, 45; and see *Inspiration* and *Insight*.
Revolt, Poetry of, 123-126, *passim*.
Reynolds, Sir Joshua, 50; as an Academician, 161, 162.
Rhapsodists, the, 9; primitive recounters, 79.
Rhetoric, and Poetry, Milton on, 27; and see 59.
Rhyme, "a memory and a hope," 54; not essential to rhythm, 56; uses of, *ib.*; "the Walkers and Barnums," 77.
Rhythm, its importance emphasized by Shelley, Watts, and others, 25, 26; Poe on, 26; the guild-mark of uttered poetry, 51-59; its universal and mysterious potency in nature and art, 51, 52; vibrations of, 52; is spontaneous, 53; method, factors, and phenomena, 54, 56; correspondence with linguistic meaning, *ib.*; Mill on, 55; poetic, distinct from that of prose, and out of place in prose, 56-59; freed by Milton, 116; splendor and melody of Shelley's and of Swinburne's, 132; dangers of excessive, 132, 133; search for new effects, 158; Whitman's and Lanier's attempts to make symphonic, 196; Cowper and the return to flexible verse, 214

Richter ("Jean Paul"), 58, 183.
Rime of the Ancient Mariner, The, Coleridge, 236, 238.
Ring and the Book, The, Browning, 192.
"Rizpah," Tennyson, 270.
Romance, Prose. See *Fiction*.
Romanticism, emotional conception of poetry, by Byron, Moore, Mill, etc., 19, 20; "The Romantic School," *ib.*, 118,— traits and leaders of, 118-127,— theory of, 262; *vs.* realism, 145, 199; the true, 151; French and German, 243; its key note, 267.
Romantic School, The, Heine, 118.
Romola, Mrs. Cross, 137.
Ronsard, 171.
Rood, O. N., scientist, 35.
"Rose Aylmer," Landor, 184.
Rossetti, Christina, 269.
Rossetti, D. G., human passion of, 269; and see 130, 131, 158, 168, 238, 283.
Rousseau, painter, 246.
Rowland, H. A., physicist, 35.
Rubáiyát, Omar Khayyám, 212; quoted, 217.
Rubinstein, A., musician, 9.
Rules, their inefficieney, 11.
Ruskin, on the nature of poetry, 20; on the "pathetic fallacy," 204 *et seq.*; cited, 142; and see 50, 58.
Ruth, The Book of, 175.

SADNESS, an effect of beauty, 267.
Sainte-Beuve, critic, 143.
Salvini, actor, 110.
Samson Agonistes, Milton, 116.

Sanborn, F. B., 22.
Sanity of genius, 284.
Sappho, genius and relics of, 87, 88; and see 262.
Satire and Irony — Dante, 114; Heine's mocking note, 127.
Saturday Review, The, on genius, 279.
"Saul," Browning, 109.
Savant, the. See *Science*.
Scarlet Letter, The, Hawthorne, 55, 137.
Schiller, on rhythm, 54; quoted, 108; and see 18, 237.
Schlegel, A. W., on Taste, 47; cited, 92; and see 26, 143.
Schools, in Art and Letters, contests between the realists and romanticists, the academicians and the impressionists, etc., 144, 145, 157, 199; how evolved, 151; and see INTRODUCTION.
Schopenhauer, on the imagination, 21; on the musician, 65; quoted, 130; and see 270.
Science, — economics, sociology, and poetry, 14; its apparent deprecation of æsthetics, *ib.*, — but only to afford a new basis, 152; as the antithesis to Poetry, 20, 21, 28; how far antithetical, 28; illustrated — by the Aurora fresco, 29, by poetry and the weather bureau, 30; the distinction one of methods, 31, 32; discovery through imagination, 32; science and religion, 33; discussion of science and poetry, in *Victorian Poets*, 33; effect on poetic diction and imagery, 34; intuitions of the savants, 35 *et seq.*; Prof. A. S. Hardy on, 36–38; inferring immortality from evolution, 37; control of public interest, 38; really an adjuvant to poetry, 38, 39; Tyndall on Emerson, 39; its beginnings suggested in Dürer's "Melencolia," 140, 141; men of, their vitality, 142; our attitude toward Nature, 207; Lucretius and the *De Rerum Natura*, 212, 217; the new learning, 220, 296; the modern speculative imagination, 228; and the future, 250; evolution, 257; and the new faith, 291.
Science of Verse, The, Lanier, 61.
Scotland, poetry of, 264.
Scott, Walter, contrasted with Morris, etc., 131, — with Arnold, 135; source of his lyrical method, 238; quoted, 181; and see 60, 121, 263.
Sculpture, the sculptor's working method, 6; specific province of, 63; how far imitated by poetry, 67; opportunity of our native sculptors, 200; Ward, Donoghue, Tilden, 200; and see 13.
"Sea-Shell Murmurs," Lee-Hamilton, 206.
Seasons, The, Thomson, 189.
Self-Consciousness. See *Subjectivity*.
Self-Expression. See *Subjectivity*.
Semitic literature. See *Bible, Poetry of the*.
Sénancour, 135.
Sensation, Human, art must be

ANALYTICAL INDEX 331

adapted to our faculties, 198, 199.
Sensibility, Poetic, 122.
Sensitiveness — is genius a neurotic disorder? 141, 142.
Sensuality, outlawed of the highest art, 262.
Sensuousness, the Miltonic canon, 27; how sensuous impressions affect the soul, 154; required by Goethe, 247; and see 262.
Sententiousness, the gift of "saying things," 213.
Sentiment, lyrical, of southern Europe, 264; of British song, 264, 265; and see *Passion.*
Sentimentalism, foreign to our analysis, 8; of the minor Byronic poets, 121; Keats on, in the preface to "Endymion," 122; and see 265.
Seriousness — the poetry of Christendom, 143.
Shah Nameh, Firdusi, 111.
Shairp, J. C., 218.
Shakespeare, his dramatic insight, 69; subjective sonnets, etc., 79, 168; wisdom of, 98; the unrevealed and "myriad-minded," 101; the playwright, stage and period, 105; and Homer, 106; *The Tempest,* 106, 107; impersonality of, 107; "not a man of letters," 151; his truth to nature, 189; his stage effects, 191; his errors of fact immaterial, 201; the preëminent exemplar of the imaginative faculty, 229–231; his future hold, 230; intelligibility of, 236; his absolutely imaginative beings, 238; not strong in construction, 238; verbal felicity of, 240; his artistic self-control, 243; and Webster, 249; method of, 273; his dramas and the stage, 274; his faith, 290; quoted, 229, 236, 274; and see 57, 76, 113, 128, 170, 172, 179, 248, 256, 283.
Shelley, his *Defense of Poetry,* 25; *The Cenci,* 69; Arnold on, 117, 118; poetry and character of, 123, 124; as Swinburne's predecessor, 132; his mission ethical, 218; false criticism of his life and works, *ib.*; his diction, 241; as the "poet of cloudland," 246; his imagination, *ib.*; quoted, 143, 266; and see 89, 90, 169, 173, 179, 208, 251, 290.
Shelley, Mary (Godwin), 246.
Shenstone, W., on poets as critics, 12.
Sidney, Sir Philip, in the *Defense of Poesie,* 23; a maxim of, 140; quoted, 62, 258; and see 57.
Sienkiewicz, novelist, 137.
Simonides, 87.
Simplicity, Milton's requirement, 27; essential to beauty, 175–177; of the Hebrew idyls, 175; of the antique and of the modern, 176; may be assumed, 177; poetic force of a direct and simple statement, 193, 194; Bryant's, 252.
Sincerity, averse to foreign and classical reproductions, 201; noble skepticism, 217; of emotion, 261.

"Single-Poem Poets," 171.
"Sir Galahad," Tennyson, 266.
"Sisters, The," Tennyson, 270.
Skepticism, the nobler, of Lucretius, Omar, Shelley, etc., 217 *et seq.;* and see *Faith.*
Smyth, A. H., quoted, 169, 170.
Snider, D. J., on the Homeric epos, 95.
Snow-Bound, Whittier, 268.
Society Drama, 275.
Society Verse, Horace and his successors, 93; and see 226.
Sohrab and Rustum, Arnold, 134, 194.
Songs and Lyrics, Jonson's, Fletcher's, Suckling's, Waller's, French chansons, etc., 170–172; endurance of, *ib.;* translation of, *ib.;* songs — as distinguished from the pure lyric, 178, — English, German, etc., 179, — martial, national, etc., 266; English song dirges, 184.
Songs before Sunrise, Swinburne, 262.
Sonnets, Milton's, 117; Rossetti's *House of Life,* 269.
Sonnets from the Portuguese, Mrs. Browning, 128.
Sophocles, on Æschylus, 46; and the Greek drama, 98, 99; impersonality of, 107; and see 142, 169, 190, 238.
Sordello, Browning, 108.
Sound. See *Rhythm* and *Melody.*
Southey, 235.
Specialists, poetic, 80.
Speech. See *Language.*
Spencer, Herbert, on music, 65.

Spenser, as a picture-maker, 67; art of, 170; and see 90, 115, 249.
Spirituality, the universal extramundane conception of beauty, 163.
Spontaneity, a test of, 11; Arnold's, 134; *vs.* the commonplace, 219; and see 135, 227, 264, 284, 285.
Stage, The, Elizabethan, 191; and see 271; also *The Drama.*
"St. Agnes," Tennyson, 266.
St. Gaudens, A., sculptor, 13.
Stendhal. See M. H. Beyle.
Sterility, metrical impotence, — want of creative power in adroit mechanicians, 44; uncreative periods, 48.
Stoddard, Elizabeth, 253; her novels, 273.
Stoddard, R. H., "An Horatian Ode," 239; his imaginative odes and blank verse, 252; quoted, 237; and see 129, 179.
Strafford, Browning, 110.
Street, A. B., 190.
Style, Browning's lack of, 91; Tennyson's, 91; charm of Vergil's, 91; Shakespeare's, 109; the Miltonic, 116; extreme individuality of Swinburne's, 132, 133; is subjective, 144; directness, 192, 193; methods as affecting quality, 214, 215; "The grand manner," 248.
Subjectivity (the personal note), one of the two ruling qualities of art, 77; the subjective poet, 80, 81; specialists, *ib.;* how far

a trait of the speechless arts, 80; of the Hebrew lyrists, 82–85; modern self-consciousness, 85; of the Greek lyrists, anthologists, idyllists, 87–90; of Euripides and his satirist, 88; of style, as in Vergil, etc., 91; of Latin song, 92–94; the cry of adolescence, 79, 101; the subjective modern drama of Browning, 108; subjective undertone of later epic masterpieces, 111,112; of Dante, 113; beginning of English self-expression, 115; of Milton's epic and sonnets, 115–118; typical of the nineteenth century, 118; finds its extreme in the Romantic and Georgian schools, 118–127; necessity for self-utterance in youth, its force and dangers, 121; Byron an eminent exemplar of, 122, 123; Shelley's, 124; Heine's, 125–127; may be termed feminine, 127; Mrs. Browning's, 128; qualified vision of the modern art-school, 129 *et seq.*; of style, in Swinburne, 131; a matter of temperament, 133; Arnold's theory against, 133, — his subjective poetry, 134, 135; not opposed to genius, 139; a pervading characteristic of the poetry of Christendom, as contrasted with the antique, 139–143; sympathetic quality, 140; its introspection typified by Dürer's "Melencolia," 140, 141; to what extent neurotic sensitiveness, 141, 142; resulting loss and gain, 142; great worth of individuality in style and feeling, 143 *et seq.*; the poet finds in nature his own mood, 202–204; the "pathetic fallacy," 204–210; of English emotional verse, 265; and see 250.

Sublimity, of the Vague, 244; and see *Imagination*.

Suckling, Sir J., 171.

Suggestiveness, in Japanese art, 31; an imaginative factor, 239; and see 59.

Supernatural, the, not always an imaginative element, 236, — but sometimes purely so, 238; of the Southern and Northern literatures, compared, 243.

Supernaturalism, of Camoëns and Milton, 245.

Swift, 94.

Swinburne, subjective style of, 132; strength and beauty of his dramas, 132; certain results of his lyrical individuality, 132, 133; his erotic verse and true passion, 262; quoted, 63; and see 131, 179, 200, 203, 251.

Symbolism. See *Allegory*.

Sympathetic quality. See *Subjectivity*.

Sympathy, poets of, 268.

Symphonic quality, — the Liturgy, 293.

TAINE, 8, 50; his theory of environment, etc., 276.

Tale of Two Cities, A, Dickens, 55, 137.

Talent, as distinguished from genius, 279; value of, *ib.*
"Talking Oak, The," Tennyson, 215.
"Tam o' Shanter," Burns, 268.
Tasso, 79, 112.
Taste, renounced by Wordsworth, 20; Poe's view of its preëminence, 26; as wrongly forsworn by certain poets, 47; Schlegel on, *ib.*; often falsely assumed, 48; "the artistic ethics of the soul," 49; discordant, how produced, 158; inborn, though cultivable, 161; and fashion, *ib.*; limitations of Anglo-Saxon, *ib.*; the maxim *De gustibus*, 166; not always allied with creative faculty, 280; an exquisite possession, 280, 281; and see 283.
Taylor, Bayard, imaginative poems of, 254; and see 129, 195.
Taylor, Sir H., *Philip van Artevelde*, 104; dramas of, 132.
Taylor, Jeremy, 59.
Technique, poetic forms, measures, etc., 77; Hebraic alliteration, parallelism, etc., 85; blank verse, 105; modern mastery of, 130, 225; "French Forms," 158; over-elaboration and over-decoration, 177.
Temperament, Hugo's, 119; Landor's, Alfieri's, etc., 125, 133; the artistic is androgynous, 127; modern view of the poetic, 141; the Greek, 142; governs a poet's product, 133; Arnold's in conflict with his theory, 133-135; should be respected, 135, 136; the English, as respects taste, 161; of Pope, 214.
Tempest, The, Shakespeare, analysis of its components, 106, 107.
Tennyson, his "Day-Dream" quoted, 68-70; as poet-painter, 68, 70; early poems of, 168; as a technicist, 177; his dramas, 191; as a poet of nature, 193; as an idyllist, *ib.*; on Truth, 198; *In Memoriam*, the representative Victorian poem, 212; sententiousness of, 213, 215; *The Princess*, 237; vocabulary of, 242; anecdote of, 255; his gain in passion, 269; quoted, 35, 102, 167, 187, 208, 264, 291; and see 10, 130, 131, 136, 142, 172, 179, 200, 225, 235, 266, 268.
Terence, 100.
Thackeray, cited, 103; and see 58, 137, 215, 283.
"Thalysia," Theocritus, 90.
"Thanatopsis," Bryant, 252.
Theatre. See *The Drama* and *Dramatic Poetry*.
Theme, not always the essential factor, 236.
Theocritus, Vergil's imitations of, 91; quoted, 179; and see 90, 193.
Theognis, 212.
Thomson, James [1834-82], 133.
Thomson, *The Seasons*, 189.
Thoreau, 193.
Thought, poetry as the voice of the conscious intellect, 48; must not disregard beauty, 48, 49; exact, inexpressible by music, 66; and see 147.

Thoughts on Poetry, Mill, 20.
Three Memorial Poems, Lowell, 267.
"Threnody," Emerson, 267.
Tilden, American sculptor, 200.
"Tintern Abbey," Wordsworth, 205.
Tolstoi, 137.
"To Mary in Heaven," Burns, 265.
"To the Sunset Breeze," Whitman, 253.
Tourgénieff, novelist, 58, 137.
Tradition, the war upon, 290.
Tragedy, Aristotle's definition, 103; ethics of, *ib.*; why it exalts the soul, 103, 271, 272; its reconciliation of Deity and Destiny, 104; and see *The Drama*.
Training, poetic, 9–11; cleverness of modern, 59–61; a true vocation indispensable, 60; "The Science of Verse," 61; self-consciousness in youth, 101; Heine's song-motive, 127; of the taste, 166; effects of town and of country in youth, 195; the formative period, 218; juvenile contemporaneousness, 226; a neophyte's errors, 235.
Transcendentalism, from Plato to Emerson, 21–24; the Concord School, 23, 24; indifference to form, beauty, etc., of certain modern idealists, 149.
Transition Periods, 114; the recent one, 294.
Translation, from the Sanscrit, Arabic, etc., 82; English version of the Hebrew Scripture, 85; renderings from the Greek anthology, by Cory, Lang, etc., 89; does not convey the full *beauty* of a poem, 166; of early French chansons, etc., 171.
Trilogy, Swinburne's, of Mary Stuart, 132.
Tristia, Ovid, 92.
Truth, the essential verities, 4; wisdom and ethics of the grand drama, 97–104; Browning's philosophy, 108; Dante as an ethical teacher, 114; *The Faerie Queene*, 114; pure, symbolized by beauty, 168; *as an element of poetry*, 187–223; what is meant by its unity with beauty, 187; the didactic heresy as the gospel of half-truths, 188; a matter of course in good art, 189; incidental, better than premeditated, *ib.*; side-glimpses of it more effective than details, 190; broad and universal, or minute and analytic, 191, 192; of Browning and Tennyson, in comparison, 192, 193; requires naturalness, 193; force of its direct statement, 193, 194; of Wordsworth and Bryant's broad method, 194; of the American poets of nature, 195, 196; not a display of mere facts, 196; nor a servile imitation, 197; is alive with interpretation, 198; must pay regard, also, to things as they appear to be, 198; of realism and ideality, 199; of fidelity to one's environment, *ib.*; of "local flavor," 200; of sincerity,

201; not to fetter the poet's imagination, 201; mere description unsatisfactory, 202; largely subjective, 202, 203; question of the "pathetic fallacy," 204–210; of nature's apparent sympathy, 205; scientific truth, the fearless desire for, 207; philosophical truth, 211–219; of the higher didacticism, 211–213; of ethical insight, 216; of a noble iconoclasm, 217–219; hostile to the commonplace, 219; free and alert, 220; finally coherent with beauty, 220, 221; the God of, also the God of art, 223; and see 46, 147, also *Ethics* and *Didacticism*.

Tudor sonneteers, 115.

TURNBULL MEMORIAL LECTURESHIP, THE PERCY, 4; its founders, 6; design of its initial course, *ib.*; theory of these lectures, 76; and see 93.

Turnbull, Percy Græme, 93; and see INTRODUCTION.

Turner, J. W. M., painter, 210, 246.

Two Worlds, Gilder, 257.

Tyndall, scientist, quoted, 39.

UNCONSCIOUS, THEORY OF THE, Hartmann, 46, 147, 156.

Universality, of the arts, 13; world-poems, 112; Shakespeare's, 230; of genius, 283; and see 191.

"Universal Prayer, The," Pope, 214.

Universities, ideal side of, 4, 5.

Utility, its relation to beauty, 156;

La Farge on the relations of fitness and beauty, 163.

Utterance, poetry is, 62; and see 264, also *Expression* and *Language*.

VAGUE, THE, imaginative effect of, 244–247; in Hebrew poetry, 244; in Camoëns, Milton, Coleridge, 244, 245; in Shelley's cloudland, 246; of thought and style, its reflex action, 235.

Vanity Fair, Thackeray, 137.

Variety, advance in poetic materials for color, diction, etc., 176.

Vates, the, 287.

Vedder, E., painter, 212.

Velasquez, painter, 150.

Vergil, the Vergilian style, 91; "Tu Marcellus eris," 93; quoted, 286; and see 43, 212.

Véron, E., critic, his subjective theory of Beauty and the Æsthetic, 152, 157; cited concerning Genius, 283.

Vers de Société. See *Society Verse*.

Verse, the true antithesis of Prose, 20.

Versifiers and Verse-making, 8, 11, 13; delusion of poetasters, 24; Milton on rhymers, etc., 27; Sidney on, 62.

Vibrations, their function, as the only media through which impressions reach the incarnate human soul, 52, — the soul thrilled by, and responsive to them, *ib.*; impalpable, 53; musical, 65; excite reflex action, 72; their expression of the quality of Beauty, 153-155; actually operative, 154; appeal, through all the senses alike, to spiritual feeling, *ib.*; Beauty's under-vibration, 180; and see *Rhythm*, etc., and INTRODUCTION.

Victorian Period, School of, 55; Browning, 110; *In Memoriam*, 212; its reserve, 265; and see 125, 138.

VICTORIAN POETS, by the author of this volume: references to, 33, 61, 108, 177, 192, 226, 269.

Villon, 167, 171, 184.

Virility, of the ancients, 142; of scientists, *ib.*; of recent poets, *ib.*; healthfulness of impersonal effort, 142; and see *Masculinity*.

Vision, absolute and unconditioned, 77-80; conditioned, 80; clearness of the artistic, 233; Blake on, *ib.*; the poet dependent on, 234; and see 255.

Vocabulary, the poet's, how acquired, 10; and see *Diction*.

Volapük, 216.

Voltaire, 82.

WALLENSTEIN, Schiller, 104.
Waller, 171.
Ward, J. Q. A., sculptor, 13, 200.
Watts, T., essay on Poetry in the *Encyc. Brit.*, 25, 26, 28.
Waverley Novels, the, 131.
Webster, John, 108; *The Duchess of Malfi*, 249.

Webster, D., and Choate, reminiscence of, 192.
"Wertherism," 121.
Westward Ho! Kingsley, 137.
"West Wind, Ode to the," Shelley, 266.
"What is the Use?" Ellsworth, 289.
White, R. G., critic, cited, 246.
"White Rose, The," anon., quoted, 171.
Whitman, W., his Americanism, 129; as a poet of Nature, 195; compared with Lanier, 196; his defects, *ib.*; genius and cosmic mood, 253; quoted, 38; and see 35, 158.
Whittier, national sentiment of, 129; passion of his song, 268; a poet of sympathy, *ib.*; *Snow-Bound, ib.*; and see 136, 195.
"Will Waterproof's Lyrical Monologue," Tennyson, 215.
Wilson, J., 58.
Winter's Tale, A, Shakespeare, 189.
Wisdom, of true genius, 284; and see *Didacticism, Truth*, etc.
Wit, 213.
Witch of Atlas, The, Shelley, 246.
With Fire and Sword, Sienkiewicz, 137.
Wonder, 245; and see *Imagination*.
"Woodnotes," Emerson, 225.
Words, the Power of. See *Language*.
Wores, T., painter, quoted, 31.
Wordsworth, on imagination, 20; on prose and verse, *ib.*; on po-

etry as the antithesis of science, *ib.*, and 28; on insight, 45; his classification of his poems, 77; his more enduring poems, 172; his pathetic lyrics, 184; broad effects of, 194; his repose, 203; his study of nature's effects upon himself, 204; reasons for his influence, 210; Shairp's view of, 218, 219; "the faculty divine," 259; on poetry as emotion, 263; the "passion" of, *ib.*; originality of, 277; quoted, 63, 136, 205, 206, 236, 261; and see 60, 125, 142, 173, 189, 190, 193, 252.

Wordsworthians, the, 47; Arnold on, 219.

Wuthering Heights, E. Brontë, 137, 273.

"Ye Mariners of England," Campbell, 266.

Zest, of the antique temperament, 139, 143; its worth, and how sustained, 160, 161; Clough's, 295.

THE END.

3

VICTORIAN POETS.

With Topical Analysis in margin, and full Analytical Index. *Twenty-first Edition.* Revised and extended, by a Supplementary Chapter, to the Fiftieth Year of the Period under Review. Crown 8vo, $2.25; half calf, $3.50.

The leading poets included in Mr. Stedman's survey are Tennyson, Landor, the Brownings, Hood, Arnold, "Barry Cornwall," Buchanan, Morris, Swinburne, and Rossetti. It also embraces very fully the minor poets and schools of the period, and with its copious notes and index forms a complete guide-book to the poetry of the Victorian era.

AMERICAN CRITICISMS.

The new chapter which Mr. Stedman has added to his "Victorian Poets" reviews the product of the past twelve years, thus bringing the English record down to even date with the "Poets of America," and making the two books more exactly the companions and complements of each other. The fresh material, which comprises about seventy pages, is devoted in a large measure to the examination of present poetical tendencies, and this is necessarily illustrated with mention of a great number of minor poets, — so many that we have a nearly exhaustive record of those entitled even to passing attention. Such a catalogue, pointed by quick touches of criticism, is of high value in defining the literary movement, and has no relation to any excessive estimate of the real value of the current poetical work. . . . We close the book with renewed admiration of the masterly handling of a fascinating but difficult subject, and with the gratification of knowing that America has produced the best book yet written on the English poetry of this age. — *New York Tribune.*

This delightful book. . . . Among the best examples of criticism in our literature. . . . We ought, in justice to its encyclopedic character, to state that it contains notices, more or less extended, of every poet of any pretensions who has flourished in England since 1835; and that these notices are not simply critical, but also biographical, those incidents and influences in the lives of the subjects which may be supposed to have had a moulding influence upon them being made prominent. The book is thus a handbook to the poets and poetry of the period. — *Hartford Courant.*

The work is as admirable in form as it is skillful in method. Biography, analysis, and criticism constitute the threefold direction in which Mr. Stedman has expended his thought. In all the field he shows himself thoroughly at home, a familiar friend as it were of the men and women of whose characters and lives he writes, an intelligent student of their writings, and an appreciative though discriminating critic of their qualities. — *The Congregationalist* (Boston).

With its companion volume, "Poets of America," it is an example of a very high if not the highest kind of criticism. While giving the best studies of individuals and schools, it yet goes beyond the individual and school, and treats the art of poetry in the most comprehensive way. It suggests to us that as modern historians have made a vast improvement in the writing of State history, in a somewhat similar way Mr. Stedman has improved on former methods of criticism. — *The New Englander.*

Like all really good work of the sort, it cannot be made known to the reader except in the author's *ipsissima verba.* We can only counsel our friends to read for themselves, and tell them that they will find store of delights in the matured and thoughtful exposition of a scholar and man of letters, who understands his business thoroughly, and is constrained neither by fear nor favor to say the thing he does not feel. — *The Churchman* (New York).

One of the most thorough, workman-like, and artistic pieces of real critical writing that we have in English. For the period covered by it, it is the most comprehensive, profound, and lucid literary exposition that has appeared in this country or elsewhere. — *Prof.* MOSES COIT TYLER, *Cornell University.*

Mr. Stedman's volume is not merely good, but it presents the best view of the poets of the present generation in England that is anywhere to be had. — ARTHUR GILMAN.

ENGLISH CRITICISMS.

We ought to be thankful to those who write with competent skill and understanding, with honesty of purpose, and with diligence and thoroughness of execution. And Mr. Stedman, having chosen to work in this line, deserves the thanks of English scholars by these qualities and by something more. He is faithful, studious, and discerning; of a sane and reasonable temper, and in the main a judicial one ; his judgment is disciplined and exercised, and his decisions, even when we cannot agree with them, are based on intelligent grounds. — *The Saturday Review* (London).

There is none among ourselves who equals him in breadth of sympathy, or in ability to resist allurement by the will-o'-the-wisp of mere form. . . . It is because Mr. Stedman strenuously endeavors to maintain his position on a truer foundation that his history of poetry in the Victorian epoch is so valuable. . . . Not only the best book of its kind, but worth (say) fifty reams of ordinary anonymous criticism of home production. — WILLIAM SHARP, in *The Academy* (London).

He has undertaken a wide subject, and has treated it with great ability and competent knowledge. — *The Spectator* (London).

The book is generous and enlightened, and bears the stamp of unfailing honesty. — *The Academy* (London).

POETS OF AMERICA.

With full Notes in margin, and careful Analytical Index.
By EDMUND CLARENCE STEDMAN, author of "Victorian Poets," etc.
Eleventh Edition. 12mo, $2.25; half calf, $3.50.

CONTENTS: Early and Recent Conditions; Growth of the American School; William Cullen Bryant; John Greenleaf Whittier; Ralph Waldo Emerson; Henry Wadsworth Longfellow; Edgar Allan Poe; Oliver Wendell Holmes; James Russell Lowell; Walt Whitman; Bayard Taylor; The Outlook.

AMERICAN CRITICISMS.

The appearance of this book is a notable event in American letters. No such thorough and conscientious study of the tendencies and qualities of our poetry has been attempted before, nor has any volume of purely literary criticism been written in this country upon so broad and noble a plan and with such ample power. . . . Mr. Stedman's work stands quite alone; it has had no predecessor, and it leaves room for no rival. — *New York Tribune.*

It is indeed refreshing to come upon a volume so devoid of the limitations of current criticism, so wholesome, so sane, so perceptive, so just, and so vivifying as we find in this collection of essays on the "Poets of America." . . . The volume may indeed be regarded as epoch-making. Its influence on our national literature is likely to be both deep and lasting. — *The Literary World* (Boston).

Mr. Stedman's temperament, training, and experience eminently fit him for the execution of a critical work on the poets of America, or, indeed, the poets of any land. He has ingrained honesty, breadth of apprehension, versatile sympathies, exact knowledge, and withal he is a poet with a poet's passion for beauty and love of song; and so he is a wise critic, a candid and luminous interpreter of the many-voiced muse. . . . The candor, sincerity, and sympathetic spirit in which Mr. Stedman treats the many themes that come under review in connection with the poets included in his scheme are apparent all through the treatise. — *The Dial* (Chicago).

Such a work involves many kinds of talent, great patience, and ample scholarship; above all, it involves genius, and if the quality of this book were to be summed up in a single word, this one pregnant word comes first to mind, and remains after fullest reflection. . . . As a body of criticism this volume stands alone in our literature, and is not likely soon to have a companion; it justifies and permanently establishes a reputation in this field already deeply grounded. It gives our criticism a standard at once exacting and catholic, and it restates, by way of commentary on our own poetry, the great underlying laws of verse. It is criticism of a kind which only poetic minds produce. — *Christian Union* (New York).

Mr. Stedman brings to the task an unusual familiarity with the whole of our literature, unusual acquaintance with the tools of the poetical guild, and a very keen notion as to how those tools have been used abroad as well as at home. . . . The studies themselves are admirable. They show a conscience which takes in good work, and, at the same time, considers the humanities, — which remembers what is due to art, and what must be granted to human frailty. — *The Critic* (New York).

The book is one which the student and lover of poetry cannot deny himself. — *Christian Register* (Boston).

It will not be possible for any sensitive reader of the poets of America to forget that Mr. Stedman is also a poet; but it will be equally impossible for such a reader to regret it. The solid qualities of the book are the result of patient, conscientious, scholarly work, which shows on almost every page; its finer qualities, the delicate touch of sympathy, the glow of hope, the spiritual magnetism, are the fruit of the poetic temperament which no amount of industry can ever cultivate unless it first has the seed. — *The New Princeton Review.*

A true critical insight enables Mr. Stedman to deal with his subject in a generous and a noble spirit, and yet in one that is eminently just and faithful to fact. His critical gifts are of a kind rarely to be found in this country, and none are more needed in our literature at the present time. — *Unitarian Review.*

This book should quickly become a standard wherever cultivated persons desire an honest, sympathetic, suggestive, entertaining, and experienced guide to the most interesting epoch of American literature. — *The Independent.*

We are greatly indebted to Mr. Stedman for this fine example of what literary criticism should be. . . . No one not himself a poet, and a poet with a noble spirit, could have written this book. — THOMAS S. HASTINGS, D. D., in *The Presbyterian Review.*

This is the history of American poetry; it is conceived and executed in the grand style of literary criticism, and it does not fall below its promise. — GEO. E. WOODBERRY, in *The Atlantic Monthly.*

FOREIGN CRITICISMS.

In his "Poets of America" Mr. Stedman displays the same competent skill, honesty of purpose, and painstaking thoroughness of execution [as in his work on "Victorian Poets"]; and he adds to these qualities the great advantage of being on his native soil. To the students of American verse his volume is almost indispensable. . . . Every one will not agree with his conclusions; but no one can differ from so well-informed and conscientious a critic without self-distrust. — *The Quarterly Review* (London).

This book, with its few and only superficial defects, and with its many solid merits, is one which most persons of taste and culture will like to possess. — *The Saturday Review* (London).

Mr. Stedman deserves thanks for having devoted his profound erudition and the high impartiality of which he is capable, to making us acquainted with the literature of poetry as it has existed from the beginning in his country. His important and thorough study is conducted with the method, the scrupulousness, the perspicacity, which he applied formerly to the work of the Victorian Poets. — *La Revue des Deux Mondes* (Paris).

www.ingramcontent.com/pod-product-compliance
Lightning Source LLC
Chambersburg PA
CBHW020234240426

43672CB00006B/521